英汉双语云南少数民族经典文化概览丛书

丛书主编◎李强

The Songs of Wine Flowing in the Bamboo Tube

竹筒里流淌的酒歌

王睿　杨亚佳　编译

云南出版集团

云南人民出版社

图书在版编目（CIP）数据

竹筒里流淌的酒歌：汉文、英文 / 王睿，杨亚佳编译. -- 昆明：云南人民出版社，2022.8
（英汉双语云南少数民族经典文化概览丛书 / 李强主编）
ISBN 978-7-222-21124-7

Ⅰ.①竹… Ⅱ.①王…②杨… Ⅲ.①少数民族—酒文化—研究—云南—汉、英 Ⅳ.①TS971.22

中国版本图书馆CIP数据核字(2022)第139647号

责任编辑：陈　晨
装帧设计：张　艳
责任校对：任建红
责任印制：窦雪松

英文校对：黄　江

英汉双语云南少数民族经典文化概览丛书

竹筒里流淌的酒歌
ZHUTONG LI LIUTANG DE JIUGE

丛书主编　李　强

王　睿　杨亚佳　编　译

出　版	云南出版集团　云南人民出版社
发　行	云南人民出版社
社　址	昆明市环城西路609号
邮　编	650034
网　址	www.ynpph.com.cn
E-mail	ynrms@sina.com
开　本	720mm×1010mm　1/16
印　张	15.5
字　数	270千
版　次	2022年8月第1版第1次印刷
印　刷	昆明德厚印刷包装有限公司
书　号	ISBN 978-7-222-21124-7
定　价	52.00元

云南人民出版社微信公众号

如需购买图书、反馈意见，请与我社联系
总编室：0871-64109126　发行部：0871-64108507　审校部：0871-64164626　印制部：0871-64191534

序　言

　　酒作为一种特殊的文化载体，在人类交往中占有独特的地位。酒文化已经渗透到人类社会生活中的各个领域，对人文生活、文学艺术、医疗卫生、工农业生产、政治经济各方面都有着巨大的影响和作用。本套系列丛书以中华民族文化自信为经，以中国少数民族经典文化国际传播为纬，全面系统地据实介绍云南少数民族社会生活特征和独特的文化传承模式。此套丛书也是我院在"十四五"开局之年，依托多语文化比较研究中心而实施的学科建设发展项目的重要组成部分。此书《竹筒里流淌的酒歌》的作者王睿老师和杨亚佳老师都是我院优秀的青年骨干教师。

　　《竹筒里流淌的酒歌》一书作为系列丛书的成果之一，在语言格式上采用英汉双语对照模式，以满足中华优秀文化经典要素国际传播的实际需求，更好地服务于国家"一带一路"建设。由于民族渊源、生存环境、历史演进、社会发展等差异，繁衍生息在云南的各少数民族，在生产方式和生活习俗方面，都有各自的风格和特色，酒文化也是各民族传统文化中的璀璨明珠。本书记述了云南少数民族的酿酒历史、酿酒工艺和技术，并从社会功能的角度，探讨了酒在维系少数民族社会关系、庆祝节日、纪念人生历程、满足精神需求等方面的作用。我们将循着这或浓或淡的酒香，去探索少数民族酒文化的内涵和特色。从一定程度上讲，此书的出版，对涉外工作人员、大专院校学生和海内外友好人士深入了解云南少数民族民俗风情，都犹如一坛陈年老酒，定会让你循着这酒香去品味民族文化人生，为此而翩翩起舞、对酒当歌、流连忘返。

　　为此而言，未能周全，当作序。

云南民族大学外国语学院多语文化比较研究中心

李瑶

2022年8月于昆明呈贡大学城

目　录

Contents

绪　言

酒的文化特质

酒既是人类物质文明的一种产物，又是人类物质文明发展的一种标志。从自然科学的角度来说，酒是一种含有机化合物乙醇（即酒精）并对人的肌体产生各种化学作用的饮料；从社会科学的角度来说，酒是一种能作用于人们精神世界并影响人们行为的物质文化现象。但与一般的物质文化现象不同，酒能渗透到人类文化的各个方面。

酒在中国的起源可以追溯到上古时期。中国甲骨文和金文中都有"酒"这个字。《诗经》中"十月获稻，为此春酒，以介眉寿"和"既醉以酒，既饱以德"等诗句，以及《史记·殷本纪》关于纣王"以酒为池，悬肉为林，为长夜之饮"的记载，都表明酒在中华大地的兴起，已有5000多年的历史。

在历史发展的进程中，中华民族不断创造出各种独特的酿造技艺，加上我国地域辽阔，地质、水资源、气候和物产等酿酒的自然条件各异，酒的品种越来越多，品质各异，显示出其文化特质的多样性。不同的酒又常来自不同的民族，都从一个特定的角度反映了不同的民族文化。可以说，在中华文明的历史上，酒文化是各民族传统文化中的重要组成部分。

尽管酒是属于饮食文化的范畴，但其所涉及的文化内容和表现出的文化意义，不仅不局限于饮食文化本身，而且已超出物质文化的范围，渗透到风俗文化和精神文化的诸多领域。在物质文化方面，各民族根据其生存环境、生产方式和生活习惯，选用不同的酿酒原料，创造了不同的酿酒工艺和技术，酿制出各不相同的酒。以酿酒原料为标准划分，有粮食酒、果酒和配制酒三大类；以酿酒工艺为标准划分，有蒸馏酒、发酵酒、配制酒三大类；以酒的度数为标准划分，有高、中、低度酒三大类。在风俗文化方面，饮酒习俗呈现出的民族文化内容更为丰富多彩。关于何时能饮、何时不能饮、如何饮、如何敬酒、某种饮用方式的意义等，各民族都形成了自己特有的习惯规约，其中所蕴含的民族文化内涵繁复而深刻。在精神文化方面，饮酒还是禁酒、酒与节日庆典、酒与人生礼仪，以及与酒相关的文学艺术和信仰祭祀活动等，从不同的角度表现出各民族的价值观、道德观、审美观和宗教观。可见，酒文化所展示出的文化内涵和意义涉及民族文化体系的各个方面。

酒文化

酒的发明、生产、流通和消费以及由此形成的酒文化，与人类生产、生活的各个方面是紧密联系在一起的，从而构成了一个与酒有关的多层次的文化网。从结构上说，酒文化是一种综合性文化。它主要表现为一种液态的物质文化，但若从酿造、饮用和社会功能的整体结构考察，它也包含了固态的物质文化、技艺文化、习俗文化、精神文化、心理文化和行为文化等。酒文化的综合性为我们通过酒文化研究其他社会文化现象创造了条件，开辟了新途径。研究者们都可通过酒文化这一窗口，在国家政治、社会经济、生产生活、历史文化、考古发掘、礼仪习俗以及婚姻家庭等领域有新发现。

酒文化具有民族性。首先，酒文化体现的是不同的民族特有的物质文化、精神文化和风俗文化。其次，不同的酿造原料、技艺、器具、饮酒方式和用酒习俗等，也都具有明显的民族性。此外，酒在各民族的现实生活中被广泛应用，但具体的内容和形式又各具民族特色。

酒文化具有渗透性。虽然酒文化只是民族传统文化中一个很小的组成部分，但却充满了活力，具有强烈的渗透性。一方面，酒在日常生活、节日、医药保健、宗教、政治、军事等各种场合都发挥着特定的作；另一方面，酒也与各种文化现象相结合，形成一些与酒有关的文化现象。如酒令是酒与诗词歌赋等文学形式的结合，酒歌、酒舞是酒与音乐舞蹈等艺术形式的结合，醉拳、醉棍则是酒与传统体育运动的结合。

酒文化是文明社会的产物。我们从酒文化的种种现象中可以寻到文明社会发展的轨迹。少数民族的酒文化，可为社会文明史的研究提供实例，如酒具。用牛角、兽骨、禽爪、畜足等制成的酒具，用各种自然材料制成的竹木碗、皮胎碗、羊皮酒褒等，至今仍在一些少数民族中使用着，都表现出独特的文化传统，对研究各民族的历史与文化都具有十分重要的价值。

酒文化与云南少数民族文化

任何一个民族文化的创造都依赖于特定的自然地理环境，人们只能在大自然这个天然舞台上从事文化创造活动，云南的少数民族也不例外。云南各民族长期以来形成的独特的生产、生活方式、各具特色的民族文化和酒文化也因其特定地域的生产、生活条件，即地形、气候等自然条件的制约，而染上了鲜明的地方色彩。

在云南这片有着千姿百态的地形地貌、多变的气候和丰富的自然资源的土地上，生活着26个世居民族。人口超过6000人并有一定聚居区域的世居少数民族有25个，包括彝族、白族、哈尼族、壮族、傣族、苗族、傈僳族、回族、拉祜族、佤族、纳西族、瑶族、藏族、景颇族、布朗族、普米族、怒族、阿昌族、德昂族、基诺族、水族、满族、蒙古族、布依族和独龙族。繁衍生息在这里的人们由于民族渊源、生存环境、历史演进、社会发展等的差异，从生产方式、生活习俗、民族节日、婚丧礼仪到宗教信仰、文学艺术和科学技术，都有各自的风格和特色。人们用不同的生态适应方式，解决衣、食、住、行等一系列问题，传承着风格各异、五彩缤纷的民族文化。

云南少数民族酿酒和饮酒的历史悠久。在战国时期"板楯蛮"①先民的酿酒技术就已经达到了相当高的水平。板楯蛮所酿造的"清酒"，是当时酒中的极品，这种古老而传统的酿酒技术至今仍是著名的日本"清酒"的主要酿造方法。《云南图经志书》记载，在三国时间，阿昌族不仅种秫为酒，而且歌舞而饮，也有多彩的酒文化。元朝初期，意大利著名旅行家马可·波罗游历云南，在其《马可·波罗游记》中多处记载了云南酿酒业的情况。徐霞客游历云南，记载了一个专门制造用于酿酒的"酒药"的村庄，说明酿酒业在明朝时已发展到相当的规模。

大多数民族都在长期的实践中探索出一整套独特的酿酒方法、工艺和技术，酿制出风味各异、种类繁多的酒，真可谓琳琅满目。例如，彝族的小锅酒、哈尼族的焖锅酒和紫米酒、傣族的树头酒、布依族的糯米甜酒、蒙古族的马奶酒、藏族的青稞酒、水族的九阡酒、苗族的泡酒和雪梨酒、怒族的咕嘟酒、傈僳族的蒸酒、拉祜族的董棕树酒、布朗族的翡翠酒、普米族的酥里玛酒和独龙族的水酒等等。饮酒也逐渐成为少数民族日常生活的重要内容。无论是农事节庆、礼仪庆典、婚嫁丧葬、宗教祭祀，还是拜亲访友、待客接物，以及烹饪美食等，酒都是必不可少之物。

酒是各少数民族维系社会关系的重要手段。在许多少数民族中，礼尚往来之"礼"，集中表现在酒上。拜访以酒为礼，迎客以酒为敬，致谢以酒示情，消仇以

---

① 板楯蛮即今苗族、瑶族祖先。《后汉书·南蛮传·板楯蛮》。

酒示诚。酒已成为人际关系的黏合剂。只要有客人到家，彝族支系白依人都要以米酒、白酒盛情招待，宾主尽兴痛饮，一醉方休，以显主人的慷慨。[①]在云南禄劝彝族聚居区，主人在亲友酒足饭饱准备离席之时还要唱《留客歌》："米酒多多有，喝了九小坛，还有九十九。"以挽留客人，表达主人对待客人的真诚情谊。两人共持一只酒碗同饮的"同心酒"，是许多民族联络感情、增进友谊、消除隔阂的主要方法。故友饮下同心酒，情感由此加深；情侣饮下同心酒，终身得以确定；朋友之间产生矛盾，喝下一碗同心酒，旧仇新怨冰消雪化。

　　酒是各少数民族节庆的主要构成部分。节日来临，人们团聚在一起，举行盛大的庆典，举起酒杯，营造喜庆的气氛，忘却了平时劳作的辛苦。酒是目瑙纵歌的必备之物。举行节日盛典的这天，景颇人身背盛满自酿美酒的竹酒筒，从四面八方聚集到一起。酒既是祭祀鬼神的重要祭品，又是参加庆典者的主要饮食。怒族鲜花节的节期长短，取决于酒的多寡，酒喝完节日便过完了。也有以酒命名的节日，如哈尼族在农历三月中旬庆祝的节日叫"喝秧酒"。喝秧酒期间，各村寨都要举办酒宴，名为"姿八夺"，大家轮流邀约喝酒。

　　酒是各少数民族人生历程的重要标志。在人生的各个阶段性仪式中，各少数民族经常把酒当作一种标志，用以象征人的生命从一个阶段步入另一阶段。酒伴随着他们走过出生、婚嫁、死亡等阶段。如布依族在头胎孩子出生时举行庆贺酒宴，称为"月米酒"。瑶族正月初一举行的"吃贺年酒"酒会上，上一年添了人口的人家，要抱上孩子，带上一壶酒、两块豆腐、一块猪肉，来向大家拜年，借此宣布自己家中增加了人口。结婚，是人生中重大事件之一。尽管各民族的婚姻礼仪各不相同，酒在从求婚、订婚到婚礼的整个过程中，是必备之物。如彝族在择偶时"吃山酒"、迎亲时喝"牛角酒"；彝族撒尼人求婚与订婚时喝"喜口酒"或"吃小酒"；傈僳族、怒族举行婚礼时喝"同心酒"；基诺族离婚仪式时喝"散婚酒"等等。办理丧事时也少不了酒。许多少数民族在亲人离世后，会请来村寨中德高望重的老人，打开粮仓，取粮酿酒招待前来参加葬礼的亲友们。在阿昌族的丧葬活动中，酒贯穿了葬礼的整个过程，为死者的灵魂送行、安葬前祭土神、埋葬时祭山神等都要用到酒。

　　酒是各少数民族表现与满足精神需求的重要媒介。在欢庆节日、新人成婚、朋

---

① 张穗、杨世全主编，大理州州志编纂委员会办公室编：白衣源流及风俗，大理方志通讯，1996 年 2 月。

友光临等喜庆时节，借酒表达欢悦的感情；亲友去世、遇到困难、心中不快时，则借酒消愁除哀。少数民族同胞的歌舞艺术则与酒息息相关。正如彝族的一首古老歌谣所唱的："彝家的礼节要喝酒，要唱歌。"云南各少数民族吟唱的酒歌已成为民间文学宝库中一颗璀璨的明珠。酒歌常用夸张、拟人、比喻等修辞方法抒情达意。其内容丰富，常描述与酒相关的生活，抒发由酒激发的情感；语言生动形象、富于生活气息；韵律抑扬顿挫、悦耳和谐。酒助歌舞之兴，歌舞酣畅之时酒兴益浓。酒与民间歌舞艺术紧密结合，把各民族的生活、风俗和艺术才干展现得淋漓尽致。

综上所述，云南各少数民族的酒文化透着不同的地域特色和迷人的民族文化光彩，将中国酒文化装扮得绚丽多姿。正如一滴水能够折射出七彩阳光一样，透过少数民族酒文化，我们可以从一个角度看到少数民族文化的全景。我们将循着这股或浓或淡的酒香，去寻觅酒中蕴含的文化珍宝，去探索少数民族文化的内涵和特色。

第一章　醇酒飘香好生活

由于特定的生存环境、社会历史背景和文化积淀的影响，酒在少数民族生产生活中占有极其重要的地位，已成为人际交往的桥梁和纽带。农事节庆、奉迎宾客、往来应酬、亲人相聚等都离不开酒。家人之间把酒话桑麻，主客之间把酒叙友情，邻里之间把酒讲和睦，青年把酒述成长，老人把酒庆福寿，可谓是"无酒不成席""无酒不酬宾"。

第一节　酒与日常生活

以酒相伴　其乐融融

云南少数民族同胞日常生活中的饮酒习俗历史悠久。据《皇朝职贡图》记载，在明初，哀牢山区的彝族群众"喜歌嗜酒"，尤其是春暖花开的正月、二月间，青年男女"携酒入山，……饮竟月，当作日而返……"[①]。明清之际，滇南景东一带的彝族有"饭可以不吃，酒不可不喝"的说法。滇西大峡谷、怒江两岸的傈僳族"喜居悬岩绝顶，垦山而种，地瘠则去之，迁徙不常。刈获多酿为酒，昼夜酖酣。数日尽之，粒食罄，遂执劲弩药矢，猎登危峰石壁，疾走如狡兔，妇从之亦然。"[②]傈僳族"嗜酒"，"醉则狂荡，男女携手顿足，吹芦笙、弹响蔑以为乐"[③]。傣族先民"嗜酒，男弹琵琶、女吹箫以为乐"[④]。纳西族在收获后，都"治衣酿酒，不计餐，

---

① 转引自《云南通志》卷一八五。

② 〔乾隆〕余庆远：《维西见闻纪》。

③ 〔乾隆〕《丽江府志略》上卷"官师路·附种人"。

④ 〔道光〕《普洱府志》卷十八"人种志"。

坐食之"①。阿昌族先民则"种秫为酒，歌舞而饮"②。

承袭已久的饮酒习俗渗透到了少数民族生活的方方面面。很多少数民族都把"喝酒"称为"吃酒"。如怒族做农活的时候吃酒，休闲的时候也吃酒；在家吃酒，出外吃酒，家里待客，更少不了酒。客人到家，不招待别样物品不觉要紧，若是不拿酒给客人吃，就是怠慢他了。再如，彝族喜欢饮酒，"有酒便是宴""无酒不成礼"。彝家以酒为尊，每当客人来到，无沏茶敬客之礼，却有倒酒敬客之俗。每逢婚嫁，彝族同胞视"酒足"为敬，"饭饱"则在其次。每当丧葬时，讲究送酒多者为最孝敬。彝族民间谚语还说"家支和解酒，姻亲喜庆酒，过年庆贺酒，虎月火节酒，羊月祭祖酒，娶妻嫁女酒，盛会洽谈酒，讨伐御敌酒，邻里闲聊酒，室内宴客酒，地边耕地酒，坡岭放牧酒，岭内岭外探亲酒"。由此可见，酒与彝族生活的各方面紧密相连。

饮酒是少数民族日常生活中的重要活动。聚居于滇西鹤庆县的彝族支系白依人，以酒为尚。亲朋相聚，所敬唯酒，甲喝乙接，轮流更替，醉倒方罢。白依妇女外出或上山劳动时，常常带上一瓶米酒，渴了就喝。白依人热情好客，无论是本民族或是其他民族的旧友新交，只要到了白依人的家中，他们都要以米酒、白酒盛情招待。宾主尽兴痛饮，一醉方休，以显主人的慷慨。③

拉祜族喜欢饮酒，男女老少皆然，群聚而饮，且每饮必歌，常常伴随着吹笙欢跳，尽兴方散。酒，在拉祜族社会生活里象征着吉祥、喜庆，在婚丧嫁娶、节庆祭典、调解纠纷、待人接物、换工互助、治疗疾病和生产劳动等社会的许多方面，发挥着沟通联系、促进友情、助人为乐、祛风止痛、消除疲劳解乏等作用。拉祜族苦聪人也是一个以酒为尚的群体。他们评判人的品质与能力的标准之一是："喝不下三碗酒算不上好汉，吃不上三块干巴算不上能人"。苦聪人的酒歌唱道：

我家的酒坛，

摆得像石堆，

密密麻麻。

我家的酒碗，

---

① 〔乾隆〕余庆远：《维西见闻录》。

② 〔明〕谢肇湖：《滇略》卷九，乾隆《腾越州志》卷十一。

③ 张穗、杨世全主编，大理州州志编纂委员会办公室编：白依源流及习俗，大理方志通迅，1996 年 2 月。

多得像鸡枞，
层层叠叠。

我家的水酒，
像泉水一样流淌；
我家的米酒，
像七里香一样香。

这样多的酒坛，
我一人抱不完；
这样多的酒碗，
我一家端不完。

这样多的米酒，
我一家喝不完；
这样多的欢乐，
我一家享不完。①

正是基于这种苦难共同承担、欢乐大家分享的群体意识，苦聪人之间洋溢着真
挚火热的信任感，哪怕外族客人来苦聪山寨，苦聪人也要倾其所有，真诚相待：

喝吧，
痛痛快快地喝！
锅搓②的心肠像米酒一样香醇；
嚼吧，
饱饱地嚼！
锅搓的心肠像火塘一样热乎。③

---

① 孙敏等编：拉祜族苦聪人民间文学集成，云南人民出版社 1990 年版。
② 锅搓：苦聪人的自称。
③ 孙敏等编：拉祜族苦聪人民间文学集成，云南人民出版社 1990 年版。

景颇族不论男女老少，不论是赶集、串亲访友、杀牛祭祀，还是婚丧节庆，筒帕①（挎包）里总是放着一个小巧的竹制"顶壶"②（小酒筒）。知己相遇，都各自拿出"顶壶"传递给对方。接"顶壶"者倒出一杯来，首先给在场的年长者喝，再给传递者喝，表示以礼相待，彼此尊重。他们普遍把酒当作一种美味的食品，没有饭菜可以，没有酒则不行。在怒族生活中，有酒喝被视为人生最幸福的事。吃罢晚饭后，大家聚到一起，饮着酒，趁着酒兴，把酒话桑麻，表达心中的情感和思想，真是莫大的快慰之事。

傣族逢年过节、迎送客人、做摆、结婚、办理丧事等都要饮酒。他们说，生活里如果没有酒，就等于歌舞没有象脚鼓。在平时的生活里，如果跨进了傣家门槛，那就要受到主人的敬酒款待，以表示欢迎和真诚。

水族把酒看得十分贵重，每逢节日和接待宾客，往往以酒肉相待。凡婚丧大事、喜庆节日、亲朋来往、上山下田、打井盖房都要饮酒。不仅男人善饮，妇女也同样能饮，特别在招待女客时，家中主妇往往热情劝酒，推杯换盏，尽兴才休。

佤族有"无酒不成礼"的习俗，因此，酒成了他们待客议事、婚娶丧葬、起房盖屋等生活大事的必备物品。可谓是家家常备酒，季季闻酒香。酒在佤族人的生活中占有重要的位置，不仅起到调节和平衡人与人之间的关系的作用，同时也给佤族人的生活带来欢乐和情趣。大多数佤族妇女都会酿制水酒。自古以来，会酿制水酒并酿制出甜蜜的水酒，是佤族人衡量妇女勤劳、贤惠的一条重要标准，也是佤族妇女们引以为豪的事。哪家的酒最好喝、最甜，就会远近闻名，而女主人就会受到人们的尊敬。

酿酒是阿昌族妇女最基本的生活技能，能否酿制出好酒成了评价姑娘才艺的重要因素。阿昌族姑娘们从小就开始学酿米酒和烧制土锅酒的方法。秋收后，青年男女结伴上山，唱着情歌，采集苦草等十八种草药配制酒曲；冬腊时节，家家户户炊烟袅袅，开始酿酒，藏之于瓮，备足供来年节日和待客之用。

酒在哈尼族、蒙古族、普米族、瑶族、独龙族等少数民族的日常生活中也是无处不见，流露出他们共同的热情好客、崇尚真诚、团结友好的精神特征。不论是登上傣家竹楼、走进彝家土掌房或普米族木屋，还是进入蒙古族的蒙古包、白族三坊

---

① 筒帕：挎包。

② 顶壶：小酒筒。

一照壁，客人们都会受到热情的招待，各民族都会按照他们的传统习俗敬上酒来。

酒入新居　喜庆吉祥

彝族自古就有盖新房饮酒的风俗，共有两次：一次是立柱子办酒，另一次是新屋大门装好后择日办"开大门"酒。届时三亲六戚拥至，主人须摆酒席招待宾客，亲朋好友也都要携带礼物前来饮酒。

怒族在开始盖房的当天晚上，主人要招待前来帮忙盖房者喝酒表示感谢。前来帮助盖房屋的人，也常带一筒酒来表示祝贺。

景颇族新房落成，要举行"进新房"仪式。进新房，先要选吉时进新火，一般是在白天。由董萨祭师拿火把先进，男主人抬铁三角一个、铁锅一口、米酒一筒和清水一筒尾随其后，女主人背一箩筐谷子后进。当晚举行群众性的大型舞蹈，唱贺新房调，以米酒祝贺主人。在这一仪式中，酒不仅起着助兴的作用，而且被视为与火、铁三角、铁锅、水和稻谷同等重要的必需品，被迎进新房，可见酒在景颇人心目中的地位。

阿昌族家庭，凡遇有上梁竖柱的喜庆日子，事先都要准备好一定数量的米酒，一是用作祭神时的"敬酒"，二是供给工匠师傅们喝，三是让前来祝贺的亲朋好友饮用。

傣族贺新房时，亲友们按时聚拢在新房附近，然后由一个身背新甑子，肩绑长刀，手捧点燃的干牛粪饼的人领头走到门口，其余亲友各带上礼品，相继来到门口。经主客双方彼此问候，主人把客人请进屋，敬上香茶和酒。客上楼后把点燃的牛粪饼放入火塘，并用点燃的火为主人家煮第一餐饭。据说，点燃的牛粪饼象征着新房永存，灶火永远不灭，今后新房主人的日子会过得像火一样旺盛。肩绑长刀表示能避邪，消灾免难，四季平安，身体健康。亲友们给主人送礼，有活鸡，因为傣族认为鸡腿能顶住两根柱子，柱子越多越牢固；有槟榔，越嚼越甜；有白线，可把主人的魂和新房拴在一起，使主人长寿，房屋牢固，五谷丰登，六畜兴旺；有炊具，祝福主人家今后会过上吃穿不愁的生活。贺新房当天，主人杀猪宰鸡，以美酒佳肴招待亲朋，客人则为主人唱贺新房的歌。

傈僳族人家在新房落成后，要择日举行隆重的进新房仪式。其中最重要的环节是请村中德高望重、子孙满堂的老人点燃新火塘里的火，而且火要烧得越旺越好，预示主人家兴旺发达、万事如意。还要用新火塘煮出的饭菜祭祖。亲友及全村的人也会带上自酿的酒前来祝贺送礼。送礼庆贺的人要成双成对，若家中人少，襁褓中的婴儿也可算为送礼的一员。大家欢聚在新房里，尽情畅饮，载歌载舞，其中唱《盖房调》必不可少。女主人和寨邻亲友对唱，在场的群众一起参加合唱，同时伴以"跳戛"舞步，进新房活动进入高潮。最后女主人唱道：

> 多谢寨头的伙伴，
> 感谢寨尾的朋友，
> 真心实意地谢谢呀，
> 诚心诚意地感激。
> 如今姑娘睡得香甜，
> 这下老妹坐得安稳，
> 为啥这样讲呀，
> 因为我有了三格新房子。
> 真难为寨头的伙伴，
> 真谢谢寨尾的朋友，
> 没有好吃的也帮盖好房，
> 没有好喝的也帮建起家。
> 从此姑娘安心做活啰，
> 从今老妹放心去劳动啰。
> 明年的这天到来时，
> 明年这月到来时，
> 再来吃小米饭吧，
> 再来喝高粱酒吧。[①]

拉祜族贺新房时有两件重要的事要做。一是要安置好家神祭祀台位，焚香点蜡，祈祷家神保佑吉祥康泰，日子越过越好；二是在屋内中柱旁安置好火塘，安放好铁三脚架，在火塘中点香祈祷。火塘和铁三脚架安放好后便不能再移动，也不准外人触摸，否则被视为不吉利。祭祀家神和安置火塘完毕后，新房的主人就准备丰盛的宴席，请前来帮忙建房的村民们畅饮。席间老人们咏唱起贺新房调，众人在新房前载歌载舞，弹弦吹笙，庆贺新居落成。

哈尼族在新房建好后，要举行"拥达达"仪式，即"贺新房"的礼俗。首先由

---

① 云南省民间文学集成编辑办公室、保山地区民间文学集成小组编：傈僳族风俗歌集成，云南民族出版社1988年版，第12～13页。

寨子里一位德高望重的长者端着一个鸡蛋和少许糯米饭率先登楼，长者之后是一群端着铁三脚架和锅盆碗筷的年轻人。人们进入楼室之后，由长者将铁三脚架支在火塘上，架柴烧火，把温暖和光明之神引进新房。从此火塘里的火就永远燃烧下去，这是这家人生命、幸运和富足的象征。点火之后，便剥开鸡蛋拌在糯米饭中，分给前来贺新房的人们品尝。仪式完毕后，房主要准备酒宴酬谢参加建盖新房的亲朋好友和全寨父老乡亲，还要请"雅习"（歌手）吟唱赞歌。哈尼人相信，主人家今后生活的好与坏，与贺新房的热闹与否大为相关。因此，新房主人会尽量把贺新房的酒席办得丰盛些，把庆祝活动安排得较为热闹和欢乐，使自己住上新房后的日子如同贺新房的日子一样欢乐，年年岁岁富足丰登。

壮族乔迁必择吉日，主人家的头件大事就是请法师诵经，并要把祖先牌位、神龛、香炉迁往新居。乔迁当天亲友云集庆贺，帮着主人家杀猪祭祖，气氛热烈，充满了对新生活的期待。亲朋入席后，大家喝交杯酒，客人向主人家祝贺人丁兴旺、老少安泰、六畜繁衍、百事和顺。客人还要唱《贺新居歌》[1]：

隔久不走这边村，
块块石板紧相连，
石板大路平如砥，
你起新屋在路边。

你起新屋方向好，
坐北朝南阴又凉，
两边石山像狮子，
前面双凤来朝阳。

走上门前第一阶，
门前品字像楼台，
门迎春夏秋冬福，
户纳东西南北财
……

贺了大厅贺正梁，
正梁红红八丈长，

---

① 梁庭望：壮族风俗志，中央民族学院出版社1987年版。

梁上雕有福禄寿，

又雕龙凤来朝阳。

崭新干栏高又宽，

子孙代代得盛昌，

儿女勤劳家兴旺，

子孙安泰人寿长。

以酒烹饪　佳肴美味

在丰富多彩的少数民族饮食文化中，酒是一种重要的配料，一直被各族人民广泛地运用于美食烹饪，并且烹饪技法多种多样。平常人家烹饪烧鱼、炖肉、炒菜和凉拌菜时，常以酒作调料。加了酒以后，菜的色香味也都更加诱人，其中最具民族特色的，就是以酒当水煮制肉食。

滇西怒江峡谷的普米族、傈僳族、怒族、独龙族群众的酒煮鸡是独具民族特色的传统佳肴，称之为"醉鸡"。以酒当水煮鸡，怒族、独龙族称之为"斜拉"，即酒肉的意思。"斜拉"的烹制办法是：以当地放养的土鸡为主料，宰杀后切成均匀的小块，先用油爆炒，再加入2斤左右陈年水酒或烧酒煮，约半小时后即可起锅享用。以这样的方式烹饪出的醉鸡，酒香和肉香合二为一，鲜美可口，是怒江峡谷的传统美食，也是年老体弱者和产妇必食的滋补名品。

滇南河谷地区的哈尼族、傣族有烹食"醉鱼"的食俗。他们把刚捕获的活鲤鱼放入水酒中，使其剧烈游动后昏醉，再以文火炖食，其味鲜美诱人。以酒当水煮食的，还有醉鸭、醉鹅。河虾则可直接用酒灌醉生食，也可以酒当水煮食。

酒经常被少数民族当作佐料，用以炒制菜肴。新鲜的肉片、肉丁、肉丝，洒上少许料酒或烧酒，猛火爆炒后，可收到肉质细嫩、清香可口的效果；新鲜牛肉整块用料酒或烧酒浸泡几分钟后，再用文火炖煮，肉鲜易烂，腥臊不存，清新爽口，许多时鲜菜在炒制起锅前淋少许的酒，鲜嫩回甜，色味俱佳，而且还有开胃消食的功效。如独龙族和怒族在油煎荷包蛋时加入少许烧酒，焖片刻即成酒焖蛋，蛋味和酒气特别浓厚。怒族在酿酒时，常把鸡蛋与刚酿好的滚烫的烧酒混合，制成蛋花酒，鸡蛋与酒充分融合，营养美味。

各民族都有以酒为辅料，腌制酸菜、酸辣椒、乳腐、酱等咸菜的传统习俗。腌制时，将主料备好，均匀地拌入草果粉、八角粉、花椒面、生姜片和辣椒面后，边装罐边洒上少许烧酒，可起到灭菌防腐的作用，又可使腌制品鲜美爽口。密封贮存一段时间后，开封取食，酸菜则酸中微带回甜，并透出丝丝缕缕的酒香；咸菜则各具特色，但都渗出酒的醇美。

酸鱼鲊是居住在滇南红河沿岸的傣族的传统美食，风味独特。酸鱼鲊的制作，首先讲究鱼的来历。农历一月间，傣族到山坡上割来茅草，扎成扇形草排，放入河水里，等鲤鱼到草排上产卵；然后把附着鱼卵的草排取回放在自家的小鱼塘里，十多天后鱼苗即孵出。早稻栽下后，再将鱼移苗入水田中，到收割季节，即可捕获到巴掌大小的鲤鱼。剖开鱼腹取出内脏，撒上盐、草果面、花椒粉，再把事先酿好的糯米白酒连糟带汁灌装进鱼腹内，之后将鱼装入罐中，密封贮存半年以上即可食用。这时，鱼肉变成淡黄色，鱼骨酥软，可生食，亦可油炸而食，酒香鱼香四溢，入口酥化。

以酒养生　健康长寿

酒的医疗保健作用，主要是通过药酒来实现的，但一般的酒也有一定医疗保健功能。适当饮酒能加快血液循环，促进新陈代谢，增强免疫能力，刺激唾液的分泌，提高消化能力，对健康不无裨益。此外，酒对于身体各部分功能逐渐衰退的老人来说，也有一定的医疗保健和延年益寿的作用，因而少数民族同胞们也把酒用于医疗保健。

例如，壮族认为，山高路陡，做工、走路都很辛苦，酒可以消除疲劳，舒筋活血，因而饮了酒身体较舒服。壮族还将鸡、鸭、猪、牛的胆汁溶于酒，据说饮后清火明目。壮族也炮制蛤蚧酒、三蛇酒、虎骨酒等名贵药酒。蛤蚧酒润肺、补肾、壮阳；三蛇酒和虎骨酒对治疗风湿、跌打损伤、腰肌劳损有效。在长期的酿、饮实践中，阿昌族积累了较丰富的经验。他们把酒运用于医疗保健，发展了多种药酒，能治疗跌打损伤、胃疼胃寒、妇女痛经、风湿等多种疾病。他们也将酒运用于日常御寒、滋补身体、防疾病和抗毒等。蒙古族的马奶酒含多种氨基酸和维生素C，可缓解因缺乏乳糖酶而造成的腹胀、肠痉挛、腹泻，并且有滋补健身、驱寒、舒筋活血、补肾消食、健胃之效。白族的糯米甜酒，是专为妇女和孕妇制作的，有滋补和催奶的作用。瑶族认为喝酒能祛湿御寒、消除疲劳、消食健胃、强健身体。佤族崇尚酒，认为酒是最高尚、最圣洁的物品，能帮助佤族人御寒、驱邪、解毒，也能养身、提神。佤族经过长期的生活实践，知道酒有很好的解毒作用。例如吃了什么山茅野菜或山菌野果中了毒，喝下几口水酒，很快就会好。在生产生活中不注意伤了皮肉，用酒洗净伤口，伤口就不会肿，而且好得快。

除了用酒炮制药物、用酒治病和用酒服药外，少数民族还将药物根块、植物果实、植物杆茎、动物的骨、胆、卵等加入自酿的烧酒中，制成各类药酒。以药入酒，以酒引药，防病治病，延年益寿。

第二节　酒与节日庆典

酒与各民族的节日密切相关。年丰岁稔，群众举杯同庆；追念先辈业绩，以酒抒怀；驱邪逐煞除秽，借酒助威；农事饮酒，以酒祈福。考察与审视各民族的节庆习俗，可以说，没有节日不用酒，没有酒不成节日。节日用酒的种种习俗，可以从一个侧面反映少数民族节日文化的特质，成为人们认识和了解相关民族的一道窗口。

驱邪除秽酒当先

酒，作为先民们最初在自然界中发现的一种普通饮料，在少数民族心目中具有特殊魔力，满足着驱邪除秽、敬神祭祖的需要。有许多少数民族节日起源于宗教庆典活动。如彝语支各民族盛行的火把节源自先民们的火崇拜意识，傣族的泼水节则始于傣族社会普遍的水崇拜观念，白族的绕山灵、纳西族的祭天节、景颇族的目脑纵歌、壮族的故密给节、拉祜族的扣木扎等都反映了少数民族广泛的祖先崇拜和自然崇拜。在这些节日中，酒被充分地神圣化了。

目瑙纵歌是景颇族规模最盛大的传统节日，它将酒的超自然力量的功能发挥得淋漓尽致。目瑙纵歌又称"总戈"，意为"欢聚歌舞"。其主要目的是消灾免难、驱邪逐煞、祈求平安、庆祝丰收和部族胜利。"目瑙"是景颇语，"纵歌"是载瓦语的直译，意思是大家一起来跳舞。目瑙纵歌节是景颇族的狂欢节，有"天堂之舞""万人狂欢舞"的美称。在德宏州景颇族聚居地，每年的正月十五前后就是目瑙纵歌节，村村寨寨都要举办目瑙纵歌。

目瑙纵歌盛典前夕，景颇山寨家家洗米、户户酿酒，竹园间飘荡着浑厚古朴的酒歌，林梢头萦绕着浓郁醉人的酒香。盛典期间，景颇族群众身背盛满自酿美酒的竹酒筒，欢聚在一起。酒既是最惹人注目的饮品，也是景颇人观念中神鬼共喜的礼物，因此大量的酒在目瑙纵歌期间被用以祭神娱鬼。每有目瑙纵歌，景颇族最重要的准备活动就是备酒。景颇族对酒的神圣力量毫不怀疑，并对之充满虔诚的情感，每有重要活动，饮用酒均有专人职掌。司酒必须是经董萨祭师认可并且德高望重者，其地位一般仅次于目瑙纵歌盛典主持者。

目瑙纵歌不仅具有强烈的文化娱乐色彩，而且已成为景颇族群众欢聚娱乐和进行物资交易的盛会。盛典期间，传统的祭祀性舞蹈目瑙舞是真正的高潮，通过象征性地祭献好鬼、逐驱恶鬼等系列活动后，参加目瑙纵歌的景颇族群众在"瑙双"（领舞者）的带领下，数万人踩着同一个鼓点起舞，规模宏大，场面壮观。舞蹈一经开场，即长达十余小时。人们饿了，以酒充饥；渴了，以酒润喉；累了，以酒提神。在充满民族特色的乐曲声中，人们尽情享受着生活的吉祥和欢愉。酒，将景颇山寨烘托渲染成一片欢乐祥和的人间乐土。

怒族朝山节期间的用酒习俗，是酒作为一种物质形式被神圣化的又一例证。朝山节因在春花烂漫的农历三月十五日举行，故又称鲜花节。传说很久以前，怒家山寨有个名叫阿茸的姑娘，不但美丽，而且心灵手巧，聪明过人。一天，她在家里织布，看见屋檐下有个蜘蛛在凌空织网，因而受到启发，发明了溜索。她和乡亲们一道，砍来金竹，编成溜索，架立在滔滔怒江之上，解决了乡亲们的渡江难题。阿茸发明了溜索被人们誉为仙女。奴隶主知道后，便派人求婚，遭到阿茸的拒绝。奴隶主恼羞成怒，派家丁来抢亲，阿茸知道后躲进深山中一个钟乳石洞。奴隶主指使家丁放火烧山，阿茸不幸遇难。怒族人民为纪念她，便把阿茸遇难的这天定为鲜花节。

贡山县怒族鲜花节的中心活动是人们带上自酿的米酒去山间的洞穴中祭祀"仙奶"。所谓"仙奶"，是被视为仙人的钟乳石上滴下的水。人们把仙奶恭而敬之地带回家，在室内环绕柱子虔诚地跳三圈舞，然后才把仙奶掺入自酿的酒中，男女老幼都要喝上一碗。喝完这碗带着"仙"味的酒，才吃饭、饮酒、歌舞。怒族相信，饮下这碗酒，仙人就会给自己的家庭、村寨带来吉祥和平安。

鲜花节这天，怒族人民一大早就起来，穿上节日盛装，带上酒和各种食物，到各自村寨附近的"仙女洞"。怒江两岸鲜花盛开，他们采来一束束鲜花，放在仙女洞周围，献给仙女阿茸；并在洞前举行祭祀，祈求吉祥如意、五谷丰登。之后，人们一家家或亲朋好友围坐在山坡上，将准备好的食物摆在铺有松针的地上，吃喝起来。他们边吃边喝边唱，兴起时又踏起欢快的舞步，山坡上充满了古朴而隆重的节日气氛。晚间，青年男女们燃起篝火，在篝火旁对唱情歌、欢快地跳舞，通宵达旦。

火把节不只是彝族的节日，还是纳西族、基诺族、拉祜族、哈尼族、傈僳族等彝语支民族以及白族的重要传统节日，不过在日子上有所不同。彝族、纳西族、基诺族在农历六月二十四举行，白族在六月二十五举行，拉祜族在六月二十举行，节期二至三天。

白族火把节起源于"火烧松明楼"的传说。传说中的主人公在一场因酒而起的灾祸中成为白族的本主，于是，白族在火把节期间，要先用美酒来祭祀火神，再敬献本主，最后才能自己享用。点燃火炬前，以酒祭祀火神和本主是火把节的主要内容。为确保火光熊熊，还要备好烈酒、松脂、食油。在举着火把走到房屋后、楼角幽室等处时，要往火把上喷酒、撒松脂和食油以助火威，从而起到驱邪除魔的作用。这样的祭祀仪式也给酒添加了神圣的色彩。

酒在彝族节庆中占据着重要地位。火把节期间，年轻的彝族姑娘们会抬着新酿的玉米酒，带着漂亮精致的酒具，到人们必经之道上布下叫"姑娘酒"的酒阵，敬献给来参加节日活动的长辈、朋友、亲戚和贵客。待摔跤、赛马、斗牛等比赛后，姑娘们会给得胜的摔跤手、彝族勇士们敬酒，唱敬酒歌，表示赞赏和敬慕。有的地方彝族青年男女"喝节酒"，每年火把节到来之前，他们就忙着为"喝节酒"做准备。姑娘买布缝制新衣、新裤、新布袋，并且买酒藏于特别的地方；男青年攒钱赶制披毡、披风，买花布和黄色雨伞，准备好银制领扣、耳环、手镯、口弦和糖果等礼物。火把节第二天，相互爱恋的男女青年们，就会找到合适的地方，三五成群地围坐在一起。女方先将酒递给各自的意中人，意中人对美酒进行一番赞美后，将酒、清水煮仔猪、烤全鸡、燕麦炒面等菜肴，摆放在女方面前，穿上情人赠送的新衣，让男女青年朋友欣赏，接受大家的夸赞。之后，男方把准备好的礼物送给自己的心上人，然后围坐一起，共享美食，谈天说地。

傣族泼水节起源于借酒降魔除恶的传说。很早以前，傣族聚居地区有个凶恶的魔王，滥施淫威，使风雨无序，庄稼难以成活。他抢劫了七个傣家少女为妻。七姐妹用计探知魔王的致命所在后，用美酒轮番把魔王灌醉，悄悄从他头上拔下一缕头发，在魔王的脖子上一拴，魔王的头颅就掉了下来。但魔头一着地即生烟起火，眼看就要祸及傣族群众的美丽家园，情急之下，大姐从地上抱起魔头，一下子烟消火灭，七姐妹就轮流着一直把魔头抱在怀里。每当轮换之际，姐妹们便互相泼清水以冲洗身上的污秽。傣族群众为了表达对七姐妹的崇敬，每年就在七姐妹杀死魔王的日子里举行大规模的泼水节。①

---

① 余嘉华主编：云南风物志，云南人民出版社1991年版，第439页。

佳节庆典酒系情

在少数民族节日活动中，饮酒、用酒习俗是增强民族凝聚力、交流思想、酝酿情绪、烘托气氛的特殊方式。饮酒，是为了欢快的情绪、热烈的氛围，是少数民族节日的主旋律。

苗族好酒。劳作之余，喝一碗以解除疲劳恢复体力；平常时节，"白酒泡苞谷饭"是苗族人最喜欢的饮食形式；闲暇时节，携一壶与知音知己者漫步花前月下，情之所至则亮开歌喉，翩然起舞；节庆时节，就更要美酒相伴了。踩花山是苗族民间一个盛大的传统节日，主要流行于滇东北、滇南的苗族村寨。各地踩花山的时间不一，但大都在春光明媚、山花烂漫的农历正月初，地点都在离村寨一定距离的林间空地。据说，苗族首领蚩尤和汉族领袖黄帝发生争战，蚩尤败北后，传令飘零四散的部落成员每年正岁初要汇聚一次，后人就把这一时间定为踩花山节。[①]踩花山最初是

苗族踩花山节

为了祭祀苗族的祖先蚩尤，后来增加的活动内容有花山祭杆仪式、爬花杆、芦笙歌舞、斗牛、武术表演等。有关踩花山节的活动，莫不与酒密切相关。节日前夕，煮粮酿酒；节日期间，饮食是酒；对歌起舞，助兴以酒；竞技活动，赏罚用酒；无不鲜明地表明酒在踩花山节中的文化功能。

踩花山当天，苗族同胞要采伐一棵高大挺拔的乔木，修枝去杈，仅留树冠，栽植在活动的主场地中间。苗族称这棵树为"花杆"，相传是由蚩尤的战旗演变而来的。他们将自家酿造的美酒盛满葫芦酒壶，高悬于花杆之冠，或放在花杆脚下。杆上拴红、黄、蓝、白四色彩带，尽显节日的喜庆。青年们从四乡八寨向花杆涌来。身着对襟短衣，头缠青色长巾，腰束布带的男子和身着镶有花边图案或挑花服装，

---

① 红河哈尼族彝族自治州民族志编写办公室编：红河哈尼族彝族自治州民族志·苗族，云南大学出版社1989年版。

佩带银质耳环、手镯、戒指和项链等首饰的妇女，伴随芦笙、唢呐、胡琴等民族乐器的拍节，围着花杆歌舞。在歌舞中，如小伙子发现意中人，就迅速解下腰间横背的雨伞，向姑娘撑去。如果姑娘不喜欢小伙，就会立即绕到姑娘圈子里躲避。若是男女双方中意，就在伞下倾吐衷情，并用对歌的形式了解对方情况。歌舞之外，是赛马、射箭、射击、爬花杆等激烈的比赛，其中，场景之壮观和气氛之热烈以斗牛为最。邱北、广南等地的斗牛活动中，组织者在赛事之前须备好美酒，谁家的牛赢了，作为奖励，主持人要向牛的主人敬酒三杯，当众一饮而尽；谁家的牛输了，作为处罚，主持人也斟酒三杯，让牛的主人当众喝下。这种赢也三杯酒、败也酒三杯的裁判原则，反映了踩花山悦人娱己的实质，也反映出苗族与人为善、待人宽厚的民族性格。

十月年是哈尼人最盛大的节日，哈尼语称之为"扎勒特"。节日前夕，哈尼村寨家家户户舂糯米粑粑，烤焖锅酒，喜待佳期。扎勒特期间，最具民族文化特色的是哈尼人称为"姿八夺"的全民性街头酒宴（长街宴），其人群之密集、场面之宏大、情绪之欢悦、气氛之热烈，鲜有能与之媲美。哈尼语"姿八夺"就是喝酒的意思。姿八夺活动当天下午，做东的村寨每户要倾其所有，发挥烹调专长，烹制一桌酒菜，抬到商定饮宴的街头或村寨某处，顺着村寨的街道一条长龙般排开。酒菜出齐，全寨老幼围坐入席，一年一度的姿八夺街头盛宴正式开始。这种全民性的长街宴，是哈尼族先民们"一人获取、众人共有"分配制度的体现，也增强了村寨民众的凝聚力。姿八夺席间，晚辈须斟酒，长者先举杯。老人们喝着自制的焖锅酒，趁着酒兴，咏唱古老的哈尼民歌。青年人纷纷向老人敬酒，当老人欣然接受后生的敬意，举杯一饮而尽时，满场齐声喝彩。人们从"龙头"席沿街吃到"龙尾"，互相祝愿事事顺心，家兴业旺，平安幸福；欢笑声回荡在古老的街巷和苍翠的林梢。

彝族年，彝语称"库史"，多在农历十月上旬择吉日而过，节期3～6天，是彝族人除旧迎新、追思先祖和祈求来年风调雨顺、五谷丰登、事事如意的重要传统节日，故而非常隆重，其中以大小凉山的彝族年极富鲜明的民族特色。

彝族年第一天清晨，青年人放鞭炮；妇女们唱着吉祥歌，舂糍粑，烤荞粑，煮鸡蛋，做米饭；男人们杀猪宰羊，准备坨坨肉等食物，并要迎接祖先回家过节。年夜饭后，中年男子成群结队到各家祝贺新年。每到一家，就吼几声，意思是告诉主人，串门的人们已经到来。进入院内，大家又吼几声，再进入主人的屋子，主人要拿出酒来招待大家，喝完后，大家又高兴地吼几声，表示感谢。如果主人又拿出好酒，人们便狂呼起来，赞美主人的大方。妇女则留在家里，招待来访的亲友和客人。晚上，家人围坐火塘，等祖宗清点人数后，陪祖宗同坐，叫"陪夜"。由最长者用荞粑、鸡蛋祭祖，随后家人喝酒、吃烤肉，有说有唱，沉浸在欢乐愉快之中。

过年第三天要祭祖送年。天亮前，各家把供在神台上的饭炒成油饭，煮过的肉再煮，新做三个或七个荞粑，再用荞面、燕麦炒面、玉米面、两把面条、一袋烟叶献祭，让祖先们趁热吃了好赶路回天上。等鸡叫头遍取下饭和肉，家人坐在火塘边，人人吃一点"送年饭"，并互祝万事如意。

彝族年节期间，小伙子身穿黑色斜襟上衣，头包帕布，右耳戴黄色或红色耳环。姑娘们穿大襟上衣，多褶长裙，衣边镶多层彩色布，头顶方帕。大家聚集一起，伴随着口琴、月琴、胡琴和芦笙的音调对歌、跳舞，或是参加荡秋千、赛马、射箭、摔跤等活动。方圆数十里的彝族人民扶老携幼赶来观看。

藏历元月初一是藏族新年节。过年前，即藏历十二月二十九日这天要打扫庭院，之后在院子里燃起一大堆火，让烟雾弥漫整个山谷。要在屋的门梁各处用土碱或糌粑画上表示吉祥的宝伞、金鱼、宝瓶、妙莲、右旋海螺、吉祥结、胜利幢、金轮等八吉祥图。还要把正月初一敬神用的供品、油炸食品、切玛（装有青稞、酥油花的五谷斗，象征吉祥）、酒、茶、人参果预备好。

初一天亮前，各家要到河边背回新年的第一桶水——吉祥水，家里的媳妇或大女儿负责熬酒。鸡叫头遍时，把熬好的"观颠"，即放有红糖和奶渣的热青稞酒端给家人喝。天快亮时，打扮漂亮的男女青年，带着风旗、神香、切玛、糌粑、青稞酒，在山顶上烧香、插风旗，祭祀山神；在地上撒些糌粑，并说"丰收喽，好丰年喽"，祈祷来年的丰收。祭祀之后，比赛谁下山快，第一个到达山脚的人，奖励三杯酒，落在最后的罚四杯。

初一早晨，也叫"初一油光光"，不但要给孩子头上抹很多油，还要把烟灰和油混合起来，把家畜的角抹得亮光光的，并喂酒糟给家畜。富裕的人家则给家畜喂新酒，在牛角尖上粘酥油花，并把家畜的旧耳饰和旧颈套换成新的。吃早饭时，阿妈把青稞酒放在座次中间，在酒壶上粘三朵酥油花。大女儿拿着酒壶，从父母开始，一杯又一杯地轮流敬酒。然后阿妈盛燕麦粥，粥里放有奶酪、人参果、肉等。喝完燕麦粥，开始喝"饭后酒"。饭后酒斟在大木碗和牛角里，要喝得一滴不剩，否则就要再喝一大碗。

从初二开始，亲朋好友彼此走访，拜年祝贺，持续三五天。客人们进门道一声"扎西德勒",主人立即迎上回敬一声"扎西德勒",有的还赠送哈达。然后宾主一起进入室内,坐在新卡垫上。主人端来糌粑,客人拈点糌粑撒向空中,祈敬天神、地神等,之后拈一点放在嘴里。紧接着,主人又提出盛满青稞酒的壶,请客人喝酒。为了尊重主人,必须三口一杯。如果喝不完,好客的主人则委托亲戚好友唱劝酒歌,歌声一落,客人须一饮而尽。

酒在藏族的其他节日中也是必不可少。射箭节是云南德钦藏族的传统节日,时间在农历四月间。节前,成年男子相聚共商过节事宜,推选一位主持人,负责筹备箭、酒等。参加者每人交箭一支、两三斤青稞,尔后酿成酒。凡男子不分老幼,均参加射箭,无力射者可由他人代射。节日开始,先举行仪式,再分成两组,射手喝酒后,入场比赛。每人可射两支箭,以中靶多少定胜负,每天反复比赛三五轮。夜晚,妇女们到靶场敬酒,为射手祝福,并燃起篝火,又歌又舞。藏族是一个十分热爱大自然的民族,他们根据高原气候、环境和生活条件,形成一种独特的民族习惯,即逛林卡。每年藏历五月一日至十五日,藏民们走出庭院,来到浓荫密布的树林,搭起帐篷,尽情享受大自然的恩赐,一边喝着青稞酒,一边尽情歌舞。雪顿节是西藏最具盛名的传统节日之一,在藏历六月二十九至七月一日之间。藏语"雪"是酸奶的意思,"顿"具有宴、吃之意。按字面解释,"雪顿"就是吃酸奶的节日。因为雪顿节期间有藏戏演出,所以又称"藏戏节"。节日期间,藏民三五成群,老少相携,在树荫下搭起色彩斑斓的帐篷,铺上卡垫、地毯,摆上青稞酒、菜肴等节日食品,边吃边谈、边舞边唱。各家还串帐篷做客。主人向客人敬酒,在劝酒时唱起不同曲调的酒歌。大家在帐篷内外相互敬酒,十分热闹。

独龙族唯一的传统节日是独龙年,农历腊月二十九为除夕,腊月三十为新年之首。独龙人称该节日为"卡雀哇"。独龙年期间,每个氏族和部落都要集体猎取野物,并将猎物分给各家各户。除夕之夜,亲友纷至沓来。客临家门,主人即用一只两耳的竹节酒杯与客人搂肩搭脖,接颊磨唇共饮同心酒。岁首清晨,曙光初照,山寨里就响起了铜锣,迎接新年的到来。早餐过后,人们随着铜锣的敲响,不约而同地来到山寨的旷地,用古朴的习俗,欢庆新年。人们不分年岁、性别、家庭,大家手牵着手,跳起本民族的传统舞蹈。长老们用编制精巧的独龙藤器,盛着可口的菜肴,以传统的方式给每个人分食。之后,歌唱声、欢呼声、舞步声便交织在一起。其后的数日,独龙人轮流在各家组织饮酒歌舞,热烈、欢快、祥和的气氛浓浓地弥散在独龙山寨。

　　"卡雀哇"期间，最重要的活动就是剽牛祭天。剽牛祭祀仪式由"纳木萨"①主持。届时，由村寨中德高望重的家族长或纳木萨把一头膘肥体壮的大公牛牵到村中广场中央，拴好立定之后，妇女们一拥而上，纷纷把珠链等饰物挂到牛角上，而后再推举出一位她们中最美丽的年轻姑娘，让她自己先披上一块色彩艳丽的独龙毯，再由她给牛披上独龙毯。待其他祭品摆好，主祭人点燃松明和松枝，口中念念有词，祈求格蒙保佑人畜平安，诸事顺利。接着纳木萨用锋利的竹矛向牛的腋下猛刺过去，牛被剽倒至死。紧接着，纳木萨要身背牛头，率众围绕"祭牛"跳舞。人们以牛为中心，自动围成圆圈，敲起铜锣，挥刀弄矛，舞蹈跳跃。然后大家煮肉分食，过年的气氛达到最高潮。大家边饮酒吃肉，边载歌载舞，共庆佳年，并祈愿来年五谷丰登，人畜兴旺。独龙江畔，变成了欢乐的海洋。最后，所有参加剽牛仪式的人都平均分得一份牛肉。

　　春节是壮族最隆重的节日，一般从正月初一到十五。年前除赶制新衣外，还要酿制甜白酒，并以酿出甜酒的好坏来预测来年的丰顺。丘北壮族，大年初一凌晨要献饭敬先祖，先由大百户献，再由各家献。献前，要送"矮兴"（送年饭）。各户将一碗米饭、三五碗菜、一壶白酒，按支系送至族长家，族长又以同样的饭菜、酒送给大百户家。当天，百户请布摩祭祖。祭毕，布摩鸣锣，令各户放畜禽。中午一二时，布摩再次鸣锣，村民闻声聚到百户家就餐。男青壮年边吃边一对对地向祖先磕头。年酒是各家族的族长备办的，族长代表族人每年向百户送年酒一罐和肉，以供全村吃年酒时用。大年初一晚上，各户长辈带着猪心和豆腐到族长家聚会。族长将自家的猪心与各家送来的猪心、豆腐放到一起，由族内厨师在族长火塘烩成一锅。煮熟后，大家按辈分入座，边喝酒边吃猪心烩豆腐，边听族长讲族史，表彰好人好事，并议定年内应做的公益事项。初二，又要在族长家中举行"更老兴"（甜酒会）。当族长吹响铜号后，族人闻声即至族长家，族长按一人一碗甜糯米酒斟上，各用芦苇管吸吮，不能用勺子舀或就碗喝，边饮边议事。族人相互有意见当面讲后就此消除，民事纠纷，是非评议，则由族长调解。

　　拉祜族过年时有拜年、饮酒和拴红线的风俗。初二这天一早，要举行盛大的拜年活动，各人要给自己的父母长辈拜年，全村人要给本村的头人、铁匠、魔巴拜年。拜年的人一般要带上一对糍粑，一瓶酒，到老人面前下跪、磕头，然后恭恭敬

① 纳木萨：独龙族的巫师。

敬地倒酒奉献。老人接过酒后欣然饮尽，高唱祝词，给他们拴红线在手腕上。之后，由头人率领全村男女老少，手举甘蔗，带上酒、肉、糍粑等礼物，到邻近村寨去集体拜年。受访的村寨必要老少倾巢出动，聚集到广场上，迎接他们的到来。接着宾主一起在芦笙的曲调中围圈而舞，跳累了退下来喝一碗酒，稍事休息再继续舞蹈，纵情狂欢，直跳到太阳落山。芦笙手和老人们沿路回到头人家中，围坐火塘，喝着烈酒饱尝肉食，欢歌达旦。也有的在第三天各自携带一些米饭和菜、一瓶酒、一点肉到铁匠家聚餐。男坐左桌，女坐右桌，尽兴欢宴。夕阳西下时，全寨又一次集合，各家献祭祖先，并将过年吃剩的秽物用火烧后丢到寨外。

达努节，又叫祝著节、祖娘节。瑶语"达努"的意思是老慈母，所以达努节也是祝寿节。相传古时候，在逶迤的群山中，有两座同样高大的山，左边的叫"布洛西"山，威武雄壮似勇士挺立；右边的叫"密洛陀"山，像个拖着长裙的姑娘。两座山每年都要互相靠近一些，经过了999年终于挨到了一起，在农历五月二十九日，随着一声惊天动地的震霹，高大英俊的布洛西和亭亭玉立的密洛陀从两山裂缝中走出来，结为夫妻。他们生有三个女儿。时间穿梭般逝去，头发花白的密洛陀遵夫嘱，让三个女儿各自去谋生。大女儿扛着犁耙，到平原耕耘，生儿育女，繁衍成汉族。二女儿挑起一担书走了，与子孙形成壮族。三女儿拿着小米、锄头到山里开荒种地，安居乐业，成为瑶族祖先。三女儿通过辛勤劳动，庄稼结出累累硕果。谁知天有不测风云，顷刻间籽粒饱满的果实被鸟兽、地鼠分食殆尽。密洛陀在女儿危难时鼓励她："天空难免出现乌云，生活也会遭受挫折，狂风吹不倒劲松，困难吓不倒勤劳的人，只要勤奋耕耘，生活是会幸福的。"并给了她一面铜鼓和一只猫。来年，庄稼长势更加喜人，她敲响母亲给的铜鼓，惊走鸟兽，放出猫吃尽了地鼠，夺得了丰收，为报祖娘养育之恩，姑娘带着丰盛的礼物于五月二十九日为母亲祝寿，共庆丰收。从此，瑶族人民将祖娘生日作为庆丰收的节日。西双版纳的瑶族于农历五月二十六日至二十九日过此节，家家户户都以四两新麻作为祭祀礼品，以表示"永不忘记"母亲的生日。节日里家家户户都要杀鸡宰羊，拿出陈年老酒，一起聚餐，出嫁的姑娘也纷纷带着孩子回娘家共庆佳节。人人都要穿上节日盛装，纵情欢庆；村村寨寨都要大摆歌台，铜鼓响起，载歌载舞，热闹非凡。

达努节最重要的活动是跳铜鼓舞。铜鼓表演需要五人出场。两人打铜鼓，一人打铜锣，一人敲皮鼓，一人舞竹帽。锣声先响，接着铜鼓、皮鼓有节奏地敲响。

铜鼓有十二套传统的打法，从不同的角度表现耕作、狩猎、与自然搏斗等场景。舞竹帽者，穿插在上述四位锣鼓手之间，不时做出幽默可笑的动作，逗得观众捧腹大笑。鼓点铿锵，舞姿纯朴，风格粗犷剽悍。入夜，灯笼、火把蜿蜒在山道上，像一条火龙向聚集点游去，人们跳起猴鼓舞、藤拐舞、猎兽舞、开山舞、丰收舞、牛角舞、芦笙舞、花伞舞等。舞罢，青年们去对歌，唱起情意绵绵的情歌。有的男女青年因对歌而订下了白头之盟。老年人则集体唱起了密洛陀颂歌，歌声充满了对密洛陀的敬意。除了唱密洛陀，他们还唱祝酒歌，每唱完一段便集体举杯畅饮、欢呼，一直到黎明仍不肯离去。

端午节时，普米族青年男女穿上节日盛装，前往深山绕岩洞。人们在岩洞的石头上点油灯，各折一根松枝或柏树枝燃烧，祈求丰收。大家把带来的米饭、肉、酒、蜂蜜、炒面聚拢会餐。大人小孩都要喝几口泡有菖蒲、雄黄的药酒，吃蜂蜜粑粑，祈求身体健康。酒足饭饱之后，普米人就到河边瀑布下洗澡、歌舞。男子持枪带犬，骑马射箭，进行围猎，兴尽始返。

农事饮酒　五谷丰登

饮酒往往也是各少数民族日常生活的实际需要。劳动之前，特别是下田或下河之前饮一点酒，有助于增加身体热量，使劳动不至伤身；劳累以后，酒能帮助解除疲乏。长期以来，酒已成为云南水族日常生活中不可缺少的常备食品，饮酒便是一种习惯。水族妇女多善酿，男子多善饮。客人来到，要是没有米酒，主妇会倍感歉意。农忙时，他们几乎每餐都喝酒。据《古州杂记》记载：这里"瘴气四时皆有，八九月尤盛，中瘴毒辄病，太阳穴痛，发热不止，眩晕呕吐……惟饮酒微醺取汗即愈。早晚酌饮醇酒数杯可以辟瘴"。[①]可见，为了避瘴和适应周围环境的实际需要，水族同胞历史上就喜爱饮酒。水族最喜爱的糯米酒，是招待客人的最佳食品。

春播献祭开秧门。春耕和播种是农业生产中的大事，很多少数民族都会在这个时节举行"开秧门"的祭祀仪式。

苗族同胞认为，秋公是教人开荒种地的始祖，当婆是铸造大地的始祖，因此，每年插秧前要向秋公、当婆献祭。苗寨人在等到秧苗已成长、可以栽插之时，就开始"开秧门"礼俗了。苗语称为"该邪介"，意思是"开始插秧"。开秧门那天，

---

① 〔嘉庆〕林溥：《古州杂记》。

在村寨头人主持下，先用鱼、肉、酒、茶泡饭向祖先献祭。然后，头人就赶到村寨公田插三丛秧苗，秧苗呈三角形。同时，立三个草标在旁边。各户人家在这天也以鸡、鸭、鱼、肉等祭祖，之后各自下田插三丛秧苗，就算是请秋公、当婆打开了天仓的"秧门"，以示开秧。祭祀的供品用毕，即被全家人聚餐食掉，之后，繁忙的插秧劳动就正式开始。

农历三月中旬，居住在哀牢山的哈尼农家的插秧季节也就到了。每逢春耕之前，各户都要选定一个好日子来过栽秧节，也叫"开秧门"。开秧门这一天，整个哈尼村寨都沉浸在一片欢乐的气氛中，他们以这种古朴的传统仪式来迎接播种的日子。一般哈尼族插秧须选定良辰吉日，属龙、猴、狗、马的日子都可以。插秧当天，家家户户都要做汤圆来敬献天地。一大早，各户男性长辈要备好糯米饭、熟鸡蛋或熟鸭蛋、米酒以及蒿枝做成的筷子，并将它们置于自家水田的水口处，行完这些仪式后方可插秧。插秧时，要先插三撮，一撮代表人吃的，一撮代表牲口吃的，一撮代表所有的庄稼，然后用竹箩罩起来，直到秧插完了之后才能拿起来。之后，围田快跑一圈，预示着今年的插秧能顺利结束。举行"开秧门"仪式的目的是希望避免天灾、避免虫灾、祈求风调雨顺、五谷丰登。按哈尼族的风俗，一家栽秧，全寨帮忙。主人将醇香甜蜜的米酒送到田里，一碗接一碗地传给正在栽秧的人们。根据风俗，"开秧门"这天，凡是从进行仪式的水田边经过的人，不论男女老少也不论是陌生人还是熟人，主人都会将你拉下田去参加这一活动。就算你不会插秧，只是随便插上几支，主人也很高兴，因为它是吉祥、幸福的象征。

每逢开春，藏族同胞都要择吉日开耕。开耕这天，他们身着盛装，端着青稞酒和切玛，庄重地来到地头。每户派出四人，一人捧哈达，一人端酒碗，一人端酒壶，还有一人专门献哈达。他们先给每家刚耕下第一犁的牛的角上抹一点酥油，再向牛轭上系一条哈达，然后给耕牛敬酒。敬酒的规矩和给人敬一样，三口一碗。用手牵住扣鼻圈，耕牛就会张开大嘴等着喝酒。如今，藏族多用手扶拖拉机耕地、播种，但他们仍先给耕牛献哈达、敬酒，还要给手扶拖拉机和耕夫敬酒。开耕之日既是牛的节日，更是人的节日。主持人按照星相师卜算的开耕方位下犁，并撒下第一把种子，即宣告春耕开始。当天，人们休息娱乐一天，次日正式开耕、播种。

庆祝丰收新谷酒。每年秋收之前，居住在云南元江一带的哈尼族，按照传统习俗，都要举行一次丰盛的"喝新谷酒"的仪式，以欢庆五谷丰登。所谓"新谷酒"，是各家从田里割回一把即将成熟的谷子，倒挂在堂屋右后方山墙上部的一块小篾笆沿边，意求家神保护庄稼。然后勒下谷粒百十粒，有的炸成谷花，有的不炸，放入酒瓶内泡酒。喝"新谷酒"选定在一个吉祥的日子，家家户户要宰大公鸡，置办丰盛的饭菜。先请村里的老人们喝，然后全家老少都无一例外地喝上几口"新谷酒"，这顿饭人人都要吃得酒醋饭饱。

农历十月初五，是云南新平苗族的爱牛节，包含犒劳赞美耕牛、庆丰收、庆团圆三个内容。这天清晨，大人忙着杀鸡、宰鸭、烧火做饭；娃娃们上山采来鲜花和红叶；老人把牛拉出来，为它除去污垢，梳理皮毛。尔后抬来糯米饭让大牛尽情饱餐。牛吃饱后，妇女们把用山花编织成的花环围在牛脖子上，男人们用红彤彤的枫叶装饰牛角。然后全家簇拥着大牛来到议牛场上。在议牛场上，老年人评议着谁家的耕牛养得好，膘肥体壮；青壮年男子互相交流犁地、耙地的技巧。议过牛后，人们目送牛群远去才回家开怀畅饮，尽情享受一年一度的团圆和丰收的乐趣。一家人在享受丰收的美味时，围绕着如何保护耕牛过冬，边吃边谈，直到夕阳西下，牛群归家。

云南兰坪、丽江地区普米族的过年时，有向过客敬酒和给牛喂酒的风俗。初一早上，普米人先在家做好饭菜，然后提着酒到村口路边等候过客，见到行人，先敬一碗黄酒，接着盛情邀请其到家中赴宴。被请到村中做客的，会感到非常荣耀自豪。客人酒足饭饱告辞时，主人要赠送食品和酒给客人。大年初二，各户带着丰盛菜肴，把牛牵到地头，给牛喂酒、烧香、磕头，象征性架着牛犁一块地，祈盼来年丰收。礼毕，大家共聚午餐。

第二章　世代传承酿美酒

云南的少数民族酿酒和饮酒的历史可以追溯到战国时期，其源远流长的酒文化还体现在各民族独特的酿造方法和品种丰富、风味独特的酒上。果酒、水酒、烧酒蒸酒、奶酒、配制酒等，不仅富有民族特色，还富有地区特点，凝聚着云南少数民族长期的、富有创造性的劳动经验和智慧。

第一节　酿酒简史

酒曲的发现与利用

随着原始农业的发展，人类开始有了剩余的粮食。然而，由于存贮条件有限，剩余食物屡屡变质，有的酸败腐坏，有的却变成可食用的自然发酵的酒。随着生产力的提高，少数民族先民们开始观察食物变酒的过程，探究并摸索促成食物发酵的途径，利用各种酒药①植物，制造出天然酒曲。根据各民族的创世史诗和神话传说，发现酒曲的人多是猎人或从事农耕的男性先祖，因此可以推测，在母系氏族晚期或父系氏族社会形成以后，少数民族就开始采集、挖掘、利用能促成食物酵化成酒的植物了。

长期流传于云南哀牢山区的彝文典籍《万事万物的开端》记载，彝族祖先色色帕尔从馊饭中悟出了酿酒的原理，但他为寻找酒药却献出了毕生的精力，最终仍没有获得成功。他的徒弟火洛尼咎继续努力，依靠集体的智慧和力量，最后才找到合成酒曲的草本原料及合成办法。史诗《万事万物的开端》②极力渲染了寻找酒曲所经历的千辛万苦：

酒是众人酿出，

---

① 酒药：云南少数民族用来合成酒曲的草本原料的总称。

② 云南民间文学集成办公室编：云南彝族歌谣集成，云南民族出版社1986年版。

色色帕尔是酿酒的祖先。

汲取九十九股清泉煮荞粒，

泉水里有九十九种鲜花的露珠，

酿酒的器具是挖空的杉树。

做酒曲要用十六种草药，

要靠几百只脚各处寻找。

火洛尼咎是做酒业的始祖，

他率领众人翻山登云，

踏出了世上的九十九条路。

荞子是众人的心血，

草药是众人的汗水，

酒曲是众人的功绩。

多少酿酒的先祖没有留下名字，

酿酒是众人的智慧。

酒曲不是一朝一夕由某一人发现的，而是在漫长的生产实践中逐步摸索总结出来的。这首史诗以极其优美的语言，创造了一种既艰辛曲折又富有诗情画意的境界，在历数发现酒药的艰难曲折时，展现了彝族先民们"劳动创造美好、劳动创造生活"的古朴思想观念。史诗虽然提出"做酒曲要用十六种草药"，但并未指明是哪些草药。流传于云南禄劝、武定一带的彝文古籍《根本·酒药歌》[1]则认为，做酒曲的草药共有十二种，并且一一指出类别，说明制作酒曲的办法：

古时酒药十二副，

六副在岩上，

岩上挖得着；

六副在山地，

山地挖得着。

---

① 云南民间文学集成办公室编：云南彝族歌谣集成，云南民族出版社 1986 年版。

岩上挖回来，

山地挖回来，

十二副草药合起来，

又舂又要筛，

水拌大麦面，

捏成小团团。

捂上七日夜，

打开晒干了，

就成为酒药。

一副"乱头发"，①

二副老黄芩，②

三副龙胆草，③

四副是柴胡，④

五副是茜草，⑤

六副一把香，⑥

七副是兰勾，⑦

八副地土瓜，⑧

九副"碎米子"，⑨

---

① "乱头发"：草本植物，有香味，状如头发，用其根。

② 老黄芩：用其根，味微甜。

③ 龙胆草：用其根，味苦。

④ 柴胡：用其叶，味辣甜。

⑤ 茜草：根红色，用其根，味辛甜，有疏经活血的功效。

⑥ 一把香：野生薄荷类，味清香，用其花。

⑦ 兰勾：草本，有助消化的功能。

⑧ 地土瓜：又名地石榴，用果，味甜。

⑨ "碎米子"：草果，用叶。

十副是提勾，①

十一副辣子面，②

十二副是草乌。③

一共十二副，

煮成好酒药。

宋朝以来，云南少数民族的酿酒业取得了长足发展，尤其水酒的民间酿制饮用已十分普遍。元初，意大利旅行家马可·波罗游滇，在其《马可·波罗游记》中，多次提到云南酿酒业的状况。这表明，宋元之际，滇中各族群众对酒曲的利用已达到了很熟练的程度。明代，民间出现了专门从事酒药配制的人。著名旅行家徐霞客漫游云南山水，沿茶马古道，由今云南楚雄彝族自治州南华县进入大理白族自治州祥云县境内，"下山过一村，北向二里，逾一坡，又二里，过一小海子，其北岗上有数家，曰酒药村"④。酒药村的地址如今已不详，但以酒曲为村名，可推测云南在明朝时就有了一定规模的酒曲商品化生产。

明清以来，随着植物医药学的发展，少数民族对用以配制酒曲的植物的认识愈加深入，许多民族已能根据酿酒的原料，通过调整酒曲中某种植物成分的比例，来酿制和调配不同品味、不同色泽的酒，以适应不同的饮用要求。如云南怒江峡谷的傈僳族以龙胆草为主要原料配制酒曲，做法是将龙胆草舂碎捏成团，蒸透，捂在竹筐中数日，发酵后即成酒曲。代代相传的傈僳族歌谣《请工调》⑤唱道：

别为蒸酒难过，

别为蒸酒发愁，

我背起了背篓，

我拎起了竹筐。

到山梁上瞧瞧，

到山坡上看看。

---

① 提勾：用叶，有香甜味。

② 辣子面：即辣椒面。

③ 草乌：用根，味苦。

④ 〔明〕徐宏祖：《徐霞客游记·卷六·滇游记五》。

⑤ 云南省民间文学集成办公室、保山地区民间文学集成小组编：傈僳族风俗歌集成，云南民族出版社1988年版。

我走过了山梁，

我走到了山坡，

见到一蓬药草，

看到一塘苦草。

找来药草交给你，

拔回苦草递给你。

你把苦草舂碎，

你把药草捏好，

苦草蒸了三天，

药草发酵七夜，

太阳出来晒晒它，

月亮出来晾晾它。

晒了三天后，

晾了三天后，

阿妹可以蒸酒啦，

老妹可以泡酒啦。

与傈僳族相同，碧罗雪山、高黎贡山一带的怒族也喜饮烈酒，并且较早就掌握了制造优质酒曲的方法。怒族配制酒曲的主料是玉米面和龙胆草。先把龙胆草捣碎，倒入冷水浸泡一天一夜。然后用泡出的红色并带有苦味的汁液与玉米面混合成面团，捏成鸡蛋大小的药团。把药团分层放在竹筐里，层与层之间再撒上米糠，以防粘连。最后把竹筐放置在火塘附近，待其发酵后取出用火烘干即可。

哈尼族焖锅酒的酒曲，采用日常所见食材如树根、树皮、树叶、果实、香料以及大米、白面等二十余种材料配制而成。为使其散发出各自独特的芳香，要精打细磨。研磨越精细，芳香越浓郁，制作的酒曲也就越好。将研磨好的酒曲原料拌入之前已经充分发酵的面中，待拌匀后捏成饼状晾干，直到需要时才拿出来使用。但是，不能一次用完所有酒曲，总要留一点来配制下一次使用的酒药，哈尼人把它称为"酒娘"。

景颇族为制酒曲还要举行采草药仪式。每年农历九月、十月间，寨中选出一对最漂亮、品行最好的小姑娘和小伙子，祭司和有威望的老人带领，背上米酒、鸡蛋、糯米饭到山上，择一空地摆上食物，坐好，由祭司唱祖先找草药的仪式歌。唱毕，一起上山采药草，带回去酿酒。他们认为，采草仪式举行得越隆重，做出来的酒药质量越好。

各少数民族制作酒曲的原料因地域差异而有所不同。拉祜族把柴胡、香树皮、香蕉皮、橘子皮、草根、带辣味的某些植物的秸秆和果实等和在一起，用铁锅炒熟、春碎，再掺入老酒药，密封后藏揞在稻草中，发酵即成酒曲。藏族的酒曲别具特色：采用一种叫"木都子格"的植物，拌以鱼、山羊、野牛等动物的胆汁，辗成粉末，再加入面粉和少许凉水捏成饼，风干即成。彝族配制酒曲用料最多，如云南禄劝、武定一带彝族，配制酒曲时常用这些原料：桂皮叶、天门冬、黄蜂、松根、萝卜、黄芩、龙胆草、薄荷、草乌、辣椒、麦芽、乱头发、马鞭梢、党参、何首乌、毛竹、天麻、草果、蜂蜜、麦面、荞面、玉米面等。但酿酒人可以根据酿酒原料、季节、自己对酒的品位和色泽的偏好等，在酒曲配料的选择、比例和配制程序上作适当的调整。哈尼族制作的焖锅酒酒曲，也因区域环境不同，在用料上有所区别，比如西北部的哈尼族喜欢多加陈皮、桂皮、茴香籽等，而西南部的喜欢多加花椒、茴香和青蒿等。

酿酒技术的发展

在战国时期板楯蛮先民的酿酒技术就已经达到了相当高的水平。《后汉书·南蛮传·板楯蛮》中记载，板楯蛮曾与秦国以清酒盟誓："秦犯夷，输黄龙一双；夷犯秦，输清酒一钟。"板楯蛮所发明和酿造的"清酒"，是当时酒中之极品，这种古老而传统的酿酒技术至今仍是著名的日本"清酒"的主要酿造方法。板楯蛮的后裔在历史演进中逐渐融入汉族、苗族、瑶族、土家族，成为今板氏族群的主流，大多分布在今云南、四川、湘西等地区。可见，战国时期云南的少数民族就已经能自酿清酒了。

隋唐时期，云南的大理一带已有较发达的酿造技术。洱海地区的河蛮人，娶妻时多事铺张，用酒达数十瓶，已有不小的生产量[①]。他们"每年十一月一日盛会客，造酒醴……三日内作乐相庆，惟务追欢"。[②]南诏时期，饮酒作乐的习俗已存在于云南各民族中，宋《太平御览》记载"磨些蛮，俗好饮酒歌舞"。但酿造时，加曲技

---

① 〔宋〕《册府元龟》卷九十。

② 〔唐〕樊绰《云南志》卷八。

术似乎不过关，樊绰说"酿酒，以稻米为麹者，酒味酸败。"[①]

宋代大理国时期，四川人杨佐来到云南买马，在距阳苴咩城150里的地方，当地的束密王以"藤觥酒"热情招待。其实，这种酒就是流行于云南少数民族地区的"钩藤酒"。南宋绍兴三年（1133年），西南蛮至四川的泸州卖马，去的人近2000人，用船带去的货物中就有酒，[②]说明从宋代起云南少数民族酿制的酒已输入内地。

元朝时，云南各民族酿酒时会加香料，以改善口感。马可·波罗游历云南时，看到昆明人"用其他谷物，加入香料，酿制成酒，清香可口"；在滇西永昌，他也看到"酒用米酿制，掺进多种香料"，称赞这是一种上等的酒品。

明《西南夷风土记》说滇南地区："茶则谷茶，酒则烧酒。"《嘉靖大理府志》收录了程本立在大理饮酒的诗："金杯哈喇吉，银筒呱鲁麻；江楼日日醉，忘却在天涯。"这里提到的"哈喇吉"，即烧酒。哈喇吉在元初忽思慧的《饮膳正要》中被称为阿拉吉酒，据考证，这是东南亚"Arrack"一词的音译，为一种用椰子做原料，经发酵蒸熘所得的蒸熘酒。这说明在嘉靖年间，云南已有较发达的蒸熘技术。"呱鲁麻"就是流行于少数民族地区的钩藤酒。明《嘉靖大理府志》记载了钩藤酒的制法："酿酒米于瓶，待熟着藤瓶中，内注熟水，下燃微火，执藤饮之，味胜常酒，名呱鲁麻。"所以，钩藤酒又称"呱鲁麻"酒，其特点是"执藤饮之"。钩藤是一种小灌木，中空可吸，饮酒时以钩藤管吸饮。《滇略·产略》有更详细的记载："钩藤，藤也，可以酿酒，土人溃米麦于瓮，熟而着藤其中，内注沸汤，下燃微炎，主客执藤以吸，按钩藤即千金藤，主治霍乱，及天行瘁气，善解诸毒，其功似与棕榈同也。"客人和主人围在酒坛边，轮流吸酒。钩藤实际上是一种冷凝管，酒在加热时开始挥发，通过藤管遇冷凝聚，说明这种酒应是蒸馏酒，即烧酒，其历史可追溯到大理国时期。现在，用钩藤饮酒的方式在滇南的苗族、壮族中仍有流行。

由于有成熟的蒸馏技术，清代云南各地出现了一些有名的烧酒。雍正年间，文献中记载了云南有"烧酒、白酒、黄酒数种"。[③]乾隆年间，《滇海虞衡志》记载了很多当时的名酒，例如力石酒。"力石酒，出定远，亦高粱烧。名力石者言其酒力

---

① 〔唐〕樊绰《云南志》卷七。

② 《资治通鉴》卷一百十二《宋高宗绍兴三年》。

③ 〔清〕雍正云南通志（卷二七·物产）。

之大，重如石也。"滇中地区元谋盆地一带出高粱酒，其味"如北方之干烧"，显然是一种蒸馏酒。同一时期，武定的花桐酒、昆明的南田酒和各地的丁香酒都很有名，而大理的鹤庆酒，"其味较汾酒尤醇厚"。

清代以后，烧酒的酿制技术在云南少数民族中迅速普及开来。从内地传入的是绍兴酒，用昆明的吴井水酿造。从缅甸传来的是古刺酒，开化（今文山）则生产洋酒，盛以琉璃瓶，应该是仿造国外的酒。未经蒸馏的酒则称为"白酒"，例如，《滇游续笔》说糯米做成甜酒，俗称白酒。现在民间也有这种"白酒"。

第二节　酿酒传说

汉文化中有酒星掌酒等关于酒的起源的传说，各少数民族也用种种传说来诠释酒的起源。这些传说可分为两类。一类是偶然发现说。如在彝族传说中，人们发现用来敬山神的野葡萄经发酵能醉人，才知道有酒这种神奇饮料的存在。[1]佤族的水酒，是外出劳动的妇女偶然将用芭蕉叶包着的冷饭挂在地头的树枝上，待其自然发酵后，人们才发现的。[2]另一类是神魔赐予说，具有神奇迷人的想象色彩。怒族认为，酒是神仙赐给人的绝妙饮料，仙人赐给怒族人民三样食品："挫确"（醋酒）、"挫辣"（烧酒）、"挫仁"（苞谷花），三者总称"挫东"，三种食品中就有两种是酒。[3]普米族的酥里玛美酒是先祖什撰何大祖冒着生命危险从妖怪那里偷学来的。[4]瑶族创世史诗《密洛陀》说，人类的始祖密洛陀是半人半神的怪物，她创造了世间万物后，开始创造人，"她拿米饭来造人，却变成了酒"。[5]在拉祜族先民的眼里，自然界万物都蕴藏醉人的酒香。在拉祜族创世神话《牡帕密帕》里，最早的酒是由天神厄莎掌管，人间万物的出现都与酒有着千丝万缕的联系：

　　……

　　冷季有了，

---

① 楚雄市民族事务委员会编：酿酒的故事，楚雄民间文学集成，楚雄市民委1988年。

② 云南省民间文学集成办公室编：水酒的来历，佤族民间故事集成，云南人民出版社1990年版。

③ 彭义良：怒族朝山节，民族调查研究1986年第1期。

④ 王震亚编：什撰何大祖，普米族民间故事，云南人民出版社1990年版。

⑤ 苏胜兴等编：密洛陀，瑶族民间故事选，上海文艺出版社1980年版。

热季有了，

果子也熟了，

可是果子没有香甜的味道。

厄莎的酒里，

好像有五种味道，

酸甜苦辣都冒着香味，

厄莎叫云雾把酒气撒到天上，

酒气变成了雨水从天而降。

雨水撒在果子上，

果子又甜又香，

果子一天天长大，

厄莎喜欢了一场。

……

景颇山寨代代相传的创世史诗《勒包斋娃》①对酒起源的咏唱别开生面：

我儿宁贯杜，

公众首领之祖啊！

为父我的吉嫩奶，

已挤在苍天与大地间；

为母我的威纯奶，

已洒在人间与世上。

以后有吉嫩奶和威纯奶洒到的地方，

会长出各种药酒植物，

酿出美酒醇又香。

为父我的吉嫩奶，

你们都能分享；

---

① 萧家成翻译整理：勒包斋娃——景颇族创世史诗，民族出版社 1992 年版。

为母我的威纯奶，

你们都会品尝。

史诗中的宁贯社，是景颇族的创世始祖，其父叫彭干吉嫩，其母叫木占威纯。宁贯杜在离开父母前经荒无人烟的世间时，聆听了父母的训诫，一是教导宁贯杜怎样寻找安身之地，二是要敬天，三是要敬地，四是指导宁贯杜酿酒，并一再重申：美酒是父母的乳汁所化。

在少数民族神话传说中，酒是天上的神仙所赐，与汉文化中"天有酒星，酒之作也"是一脉相承的。人类的先祖冒着生命危险从妖怪口中套出造酒术的传说，反映了少数民族先民们与险恶的自然环境抗争的艰难历程（普米族）；神最初造出了米饭，再用米饭造人，人还没有造出，已先造出了酒（瑶族），这是以稻作文化为生存发展背景的人们对自我历史的回顾；人类的始祖已经知道，父母乳汁化成的酿酒植物生长在崇山峻岭中（景颇族）；甚至世间万物也是因天神撒下的酒气才有了活泼的灵性和诱人的芳香甘甜（拉祜族）。透过原始文化的光环，我们可以感觉到，步入人类文明初期的先民们对发现酒由衷的喜悦之情。

这些关于酒起源的神话、传说和史诗是少数民族酒文化的重要组成部分。仔细品读，我们可以看到女性在酒文化中的重要作用，即最早发现酒和最早从事酿酒活动的都是女性。其一，各民族关于酒的诸多神话的传说中，发现酒的人有两种，一是妇女，二是猎人，其中，妇女发现的传说占绝大多数。怒族的酒是仙人赐的，其膜拜的仙人，是女性形象的钟乳石；景颇族的酒，是人类始祖宁贯杜的母亲乳汁所化。此外，彝族、傈僳族、普米族、佤族等许多民族都有通过剩饭变质而发现酒的传说。在这类传说中，发现剩饭变酒的人几乎全是妇女。

其二，在少数民族有关酿酒起源的创世史诗、传说中，最早从事这一工作的人都是妇女。"她拿米饭来造人，却变成了酒"的瑶族先祖密洛陀，是氏族女性首领的形象；佤族将酒的起源列入创世神话，神话《司岗里》认定，佤族是向蜜蜂学会酿酒技术的，而第一位能酿酒的人是佤族社会的女首领牙董。[①]流传于云南红河两岸的哈尼族迁徙史诗《哈尼阿培聪坡坡》，全面系统地咏唱了哈尼人的发展历程，史诗在叙述人类起源并学会用火之后，回忆先民们由围猎野兽到驯养家畜并走向农耕文明的漫长历程，而引导哈尼人摆脱蒙昧并走向农耕文明的人，就是一个名叫遮

---

① 艾获、诗恩编：佤族民间故事，云南人民出版社1990年版。

努的妇女。在农耕文明的基础上，哈尼族有了"吃不尽的米粮"，就"用五谷酿出美酒"。

> 哈尼还有一位能人，
> 遮努的名声飞遍八方。
> 她摘采饱满的草籽，
> 种进最松最黑的土壤；
> 姑娘又去背来湖水，
> 像雨神把水泼洒在籽上。
> 草籽发出了粗壮的芽，
> 草籽长出了高高的杆，
> 黄生生的果实结满了草秆。
> 先祖们吃着喷香的草籽，
> 起名叫玉米、谷子和高粱。
>
> 遮努的收成有好有差，
> 细想想是节令没有合上。
> 她请来放牛放羊的遮姒，
> 遮姒姑娘有了主张：
> 她指着十二种动物，
> 定下了年月属相，
> 一年分作十二个月，
> 一月有三十个白天夜晚。
> 哈尼算日子以鼠头起，
> 算到胖猪一轮就满。
> 有了属相按时栽种，
> 遮努种出了吃不尽的米粮。
> 她又用五谷酿出美酒，
> 美酒成了哈尼离不开的伙伴。①

---

① 云南省少数民族古籍整理出版规划办公室编：哈尼阿培聪坡坡，云南民族出版社1986年版。

其三，在现当代许多民族中，从事酿酒劳动的人是妇女。在景颇山寨，酿酒是妇女最基本的生活技能，景颇族女性从小就要学习酿制水酒、烧酒的方法，能否酿制出好酒，是评价一个女性生存能力、劳动技巧的重要标准。景颇族青年男女婚礼后的次日，新娘最要紧的事就是要酿制美酒，敬事公婆，要是做不出好酒，会被人终身笑话；普米族酒歌《酥里玛》调中，多次重复"阿妈阿姐酿酒忙，阿公阿爹酒瘾浓"之词；傈僳族青年在寻找意中人时，要求未来的妻子心地善良，勤劳能干，使新的家庭"早上有煮饭的人，晚上有泡酒的人"；在佤族婚礼上，老人祝福新婚夫妇"愿你们生女煮酒，愿你们生女煮饭；愿你们生男犁地，愿你们生男犁田"，这是尚未出生即已明确分工，女性酿酒的劳动必然性已经确定。

第三节　酒香四溢

少数民族先民们最先发现、最早饮用的酒是果酒。其后几千年的历史中，果酒香飘不绝。随着农业生产的不断发展，以粮食为原料酿制的酒类走入人们的生活中，其中，水酒以其悠久的历史和深远的影响，成为少数民族酒文化中最绚丽的乐章。在长期的酿制、饮用中，人们对酒的性能、功效逐渐认识，并加以利用，创造出各种有药用保健作用的药酒，为丰富生活内容、提高人们的身体健康水平发挥了独特的作用。

果酒及酿制方法

树头酒

早在元、明之际，在云南的西双版纳、德宏等热带、亚热带森林中，少数民族"甚善水，嗜酒。其地有树，状若棕，树之稍有如竿者八九茎，人以刀去其尖，缚瓢于上，过一霄则有酒一瓢，香而且甘，饮之辄醉。其酒经宿必酸，炼为烧酒，能饮者可一盏"。[1]明朝天启年间刘文征所纂《滇志·卷三十二·补物产》对树头酒也做过考证和记述：西双版纳等地"土人以曲纳罐中，以索悬罐于实下，倒其实，取汁流于罐以为酒，名曰树头酒；或不用曲，惟取其汁熬为白糖，其叶即贝，写缅书用之"。清初，树头酒就果实直接取汁酿制的方法还常见于权威性的官方文献中，清康熙《云南通志·土司》中有如下记述："土人以曲纳罐中，以索悬罐于实下，

---

① 〔明〕钱古训《百夷传》。

划实取汁，流于罐，以为酒，名曰树头酒。"据《新纂云南通志·物产考》考证，树头酒的树种，属热带椰子之类，因其果汁内含糖质，可即用于酿酒。这种不用摘取果实，而是将酒曲放在瓢、罐、壶之类的容器中，悬挂在果实下，把果实划开或者钻孔取汁酿酒的方法，着实令人大开眼界，连见多识广的江南才子檀萃历滇时，也对这种浑然天成的酿酒法惊叹不已，并在其《滇海虞衡志》的《志酒》《志草木》中列目详记。清末民初，树头取酒的办法仍残存于滇西、滇南少数民族之中，现已不可多见。

葡萄酒

《滇海虞衡志》说："滇产葡萄佳，不知酿酒，而中甸地接西藏，藏人多居之，酒盖自彼处来也。"说明早在清代中期，滇西北中甸的藏族已有葡萄酒生产。到晚清以后，法国传教士进入滇西北的德钦，将法国酿酒葡萄品种玫瑰蜜和酿酒技术带到了雪域高原，丰富了当地的葡萄酒工艺，其制作方法在滇西北的德钦藏族地区一直保留了下来。如今，在德钦茨中乡，当年传教士种下的葡萄树虽然已有上百年的历史，但仍然枝繁叶茂，果实累累。这种用玫瑰蜜葡萄和法国酿酒技术酿制的葡萄酒，逐步发展成为具有独特魅力的"云南红"葡萄酒。

此外，早在清朝时期，云南就已有桑葚酒、刺梨酒、山楂酒、葡萄酒、雪梨酒等不同果酒。如今这些果酒都已成为节庆典礼和走亲访友的必备佳品。

水酒及酿制方法

水酒，即发酵酒，用黍、稷、麦、稻等为原料加酒曲发酵而成，汁和滓同时食用，即古人所说的"醪"。水酒是云南少数民族酒中品种最多、饮用最为普遍的一类。如壮族、阿昌族的"甜酒"、哈尼族的"紫米酒"、瑶族的"糟酒"、藏族的"青稞酒"、纳西族的"窨酒"、普米族的"酥里玛"等均属此类。在许多少数民族地区，发酵酒又称为白酒，并按发酵程度的不同，分为甜白酒和辣白酒两类。

甜白酒是以大米、玉米、粟等粮食为原料酿制的。用清水浸泡或煮熟，再蒸透后，放在不渗水的盆、罐、桶等容器中，待其凉透，撒上甜酒曲，淋少许凉水，搅拌均匀，放置在温暖干燥处。夏季，1~2天即可成甜白酒；冬季，约需3~5天。拉祜族用糯米为原料酿制甜白酒。其方法是，用热水浸泡糯米和米糠，再煮沸，取出后趁热用木甑蒸透，装在陶罐内，撒上自制酒曲，放置约1~2天后即可饮用，其味清凉甜美。甜白酒实质上是在粮食中的淀粉完全糖化、而酒化过程即将开始时形成的水酒，甘甜可口，只隐约透出酒的醇香，是老幼咸宜的饮料。

各民族酿制甜白酒的历史悠久，早在元、明之际，已有商品化生产。明初，徐霞客由云南大理入永昌（今保山）途中，穿越一山峡，"有数家当南峡，是为弯子桥，有卖浆者，连糟而啜之，即余地之酒酿也"[1]。可见，早在明代，即使深山幽谷，甜白酒也成为商品，供山峡古道上匆匆过往的商旅"连糟而啜之"。甜白酒具有很高的营养价值。甜白酒煮鸡蛋，是少数民族的待客佳品，也是产妇用来滋补身体、恢复元气和催奶的保健型食品。明清以来，相袭成俗。檀萃于清乾隆间任禄劝知事，对此留

甜白酒

下很深的印象，该地区群众"每岁腊中，人家各酿白酒。开年客至，必供白酒煮鸡蛋满碗，乃为亲密"[2]。时至今日，每逢佳节良辰，泡米蒸饭酿甜白酒仍是许多少数民族最要紧的节前准备工作之一。

辣白酒是以大米、糯米、玉米、大麦、小麦、青稞、粟等粮食为主要原料酿成的低度原汁酒，属黄酒类。各族群众酿制水酒的历史悠久，明朝中后期的文献典籍中，已有辣白酒的酒曲配制与酿造的记载。

彝族辣白酒

彝族人民善于酿制辣白酒。糯米为首选原料，大米、玉米、高粱、粟等粮食也可用于酿制。酿制辣白酒的基本步骤是：

（1）浸泡或煮熟原料：将用以酿酒的原料粮用清水浸泡透心或煮熟。

① 〔明〕徐宏祖：《徐霞客游记·滇游记之九》。
② 〔乾隆〕檀萃：《滇海虞衡志·志酒》。

（2）蒸饭：将浸泡透心或煮熟的原料粮装在木制或竹制甑子内用猛火蒸透。这时的原料粮称为酒饭。

（3）凉饭：酒饭蒸透后出甑，放在干净的竹席或竹箕上，摊开，使酒饭自然降温变凉。

（4）撒曲装罐：酒饭变凉后，撒上酒曲，再淋少许凉开水，搅拌均匀即可装入清洗晾干的罐中。酒曲以自己挖掘采集植物配制的土酒曲为佳。装罐时，可直接入罐盛装，也可在罐底放置竹筛或其他竹编滤器，使罐底留出一定的空间，以分开酒糟和酒汁，使酒液清爽。

（5）出窝：酒饭入罐后，1~2天完成粮食中淀粉的糖化，形成甜白酒；5~7天后，由于酒曲中酵母菌的作用，完成酵化，酒香浓郁的辣白酒即告酿成，这时即可取出饮用或贮存了。由于酒饭装罐后要保持一定的温度以利于发酵，常将酒罐放在靠近火塘的地方，或是埋在米糠内。严冬时节，甚至用棉被来裹捂，所以，这种酿造白酒的过程也叫"捂白酒"。白酒酿成后，从糠箩或棉被中取出开罐饮用或贮藏，叫"出窝"。

（6）贮藏：白酒"出窝"后即可饮用，饮用的办法有两种，一是原汁取饮，二是根据酒汁浓度或口味需要，兑入适量的凉开水饮用。若暂不饮用，即行贮藏，可原糟原汁贮藏，也可以滤糟贮汁，或是兑入适量的凉开水贮藏。贮藏的方法是，把辣白酒取出，装入洁净的陶罐中，用稻草或麦秸秆编织的罐塞塞紧罐口，再用草灰制成的稀糊裹紧罐塞，以避免透气。用这种方法贮藏水酒，夏天可保存20天左右，冬天贮存可长达数年，贮存时间越长，酒味越是醇厚，酒劲越加绵长。所谓"彝家老酒"，就是这类长期贮藏的水酒。

积年贮藏的水酒，取出后酒香扑鼻，糟与汁已完全分离。糟浮酒面，已薄如蝉翼；酒液清澈亮丽，略呈黄褐色。饮用时醇香爽口，绝无挂喉刺鼻的感觉，饮用后神清气爽，酒劲悠然绵长，善饮者也常常不胜其力，三杯两盏之后往往就酣然而卧。敬上一碗积年贮藏的辣白酒，是彝族接待长辈尊者和佳朋良友的最高礼节之一。

纳西窨酒

纳西族先民爱饮酒。东巴经书《耳子命（饮食的一历）》，是一部农业生产劳动的颂歌，长诗的第一部分，就描写了种麦、酿酒的全过程。说明纳西族在古代就会酿酒。窨酒是滇西纳西族独创的地方水酒，在清朝道光年间开始酿制。窨酒以滇西所产的优质稻米为原料，按传统的发酵法酿制。泡米、蒸饭、凉饭后，授曲，使酒饭糖化。装罐后，低温发酵约一个月，即可取出榨酒，分缸澄清后，再密封贮存，陈酿即成。窨酒呈琥珀色，透明，味甘醇香，富含葡萄糖，多种氨基酸和维生素，有良好的滋补作用。

哈尼紫米酒

滇南谷地、红河两岸的哈尼族以当地所产的优质紫米发酵酿造而成的紫米酒，是接待宾客的最佳饮品，早在清末民初就已远近闻名。此外，种植紫米的傣族、彝族、景颇族也有酿制紫米酒的传统，酿制方法与彝族水酒相同。

傈僳族"拉酒"

云南怒江峡谷的傈僳族酿制和饮用的水酒"拉酒"，因饮用方法而得名，独具特色。傈僳族以小麦、玉米、高粱等为原料，煮熟蒸透后，拌上酒曲，密封贮存在瓦罐中发酵。嘉宾临门时，取出适量的发酵酒渣，放在锅中，置于火塘上，饮者团团围坐，主人不断往锅中加水，一面拉滤酒渣，一面斟酒敬客，直到酒味淡时为止。

独龙族竹筒酒

每当收获季节到来时，独龙族家家户户都要酿酒。他们不是用陶坛，而是用竹筒酿造。酿造时首先要选山中最好的竹子制成酒筒，再将煮熟的大米、小麦或高粱拌上酒曲装进竹筒内，七天后变得香味扑鼻，就可以饮用了。这种酒香甜可口，还有淡淡的竹子香味。

佤族"布来隆"

滇西阿佤山区的佤族称水酒为"布来隆"，是"年酒"的意思。水酒是佤族民间一种传统的散热、解毒、驱乏、壮身的清凉饮料，主要以红米、玉米、高粱、小麦、荞麦、糯谷等谷物为原料，其中，以红米、高粱为最佳。水酒的酿制方法是：将原料炒黄、磨碎，洒上少量清水后拌匀，然后用甑子蒸熟；晾凉后加入适量酒曲，拌均匀后装入铺垫有芭蕉叶的箩筐，放在火塘边待其发酵；之后将其装入备好的酒坛里，密封数月。佤族人很早就知道，酒装的时间越长，就越香甜醇和。佤族的布来隆，酿制装罐后时间不足年，是不轻易拿出来的。饮用时加入山泉水，泡上10小时左右，即成水酒，这也正是佤族把酿制水酒称为"泡酒"的原因，泡滤之后便可饮用了。

独龙水酒

滇西北独龙江流域的独龙族嗜饮水酒，其酿制方法也别开生面。独龙人酿制水酒多用玉米，也可用大米和高粱。他们将原料粮磨碎煮熟或蒸透后，晾凉，拌上酒曲。在地上挖掘一个罐形的土窖，窖的底部和四壁用干净肥硕的芭蕉叶铺垫。将酒饭放在窖内，再层层盖上芭蕉叶，使酒饭与土层完全隔离。用稀泥封糊窖口，在窖口上燃一堆火，使酒窖内的酒饭发酵3~4天后，取出已发酵的酒饭，倒入陶罐中封存若干日，即成可饮之酒。加水搅拌并加热后便可饮用，或兑上清凉的山泉水，饮之甘美醇香，消暑解渴。

普米"酥里玛"

酥里玛酒是滇西北普米族的传统佳酿，为敬神祭祖驱邪等活动所必备，也是待客访友的重要礼品。酥里玛以优质大麦为主要原料。酿造时，先将煮熟的大麦取出晾冷，再按一定的比例撒上酒曲，搅拌均匀后放入竹箩，待发酵3天后放入陶罐内，密封至少约半个月。密封的时间越长，酥里玛的色、香、味愈佳。一般3~4个月开封时，罐内已充满了酒浆，酒香四溢。开封后加入山间冰凉的清泉水，再次密封3天左右，滤出的就是酥里玛酒了。普米人就在酒浆中加入甘甜的泉水，浸泡1~2天后便可饮用。第一次加水浸泡的酒称为"头酒"，是用来敬献给远道而来的贵客饮用；第二次加水浸泡的酒称为"二酒"，普米人家在耕种、砍伐、乔迁时都用酥里玛酒配猪膘肉招待前来帮忙的人们，起到缓解疲乏、减轻劳累、营造气氛的作用；第三次浸泡的酥里玛才是主人家平常劳作解渴之用。酒糟可以用来喂养准备过年宰杀的肥猪，味道鲜美的腊肉和猪膘肉配上酥里玛，是过年时的美味。普米人自制的酒曲里有雪山上取来的龙胆草和江边的百合草，所以酥里玛酒浓香醇厚，甘冽净爽，具有独特的风味。

苗族米酒

滇东南的苗族以大米或糯米酿制水酒，方法与彝族水酒基本相同。苗族米酒含糖量高，酒精度低，是解除疲劳、清心提神的最佳饮料。苗族群众常用以佐餐，"白酒泡苞谷饭"是滇东南苗族的传统饮食。

瑶族栗树籽酒

瑶族擅长于酿制栗树籽酒。金秋时节，采下栗树籽去壳，洗净煮熟，取出滤干水分，冷却至30℃左右，加入酒曲拌匀，入坛密封两个月后即成。栗树籽酒醇香扑鼻，苦涩甜辣味俱全。

藏族青稞酒

青稞酒是藏族的传统饮料，度数很低，制作方法颇为简单。先把青稞洗净煮熟，待温度稍降，便加上酒曲，用陶罐或木桶装好密封，使其发酵2～3天后，加入清水再密封，两天后即成青稞酒。青稞酒呈淡黄色，味微酸。埋藏3～5年的陈酒，呈蜜状，饮之味浓，香气袭人。青稞酒能解渴提神，是藏族人家婚丧嫁娶、节庆典礼必备佳酿。

布朗翡翠酒

翡翠酒是布朗族以糯米为原料酿造的水酒，其制作方法与其他民族酿造水酒的方法大体相同。所不同的是，糯米发酵成酒后，布朗族在出酒时用一种叫"悬钩子"的植物叶片将糟与汁滤开，酒色透明清亮，呈翡翠色，是布朗山寨接待亲朋好友的上等饮料。

满族糜子酒

"糜"即黄米。《魏书》《隋书》《契丹国志》等书中，均有满族的先人"酿糜为酒"的记载。其做法是：将黄米用水浸泡之后上锅蒸熟，装入坛中，之后加黄酒曲搅拌均匀，两天后即可饮用。有的满族同胞将黄米炒煳后，以同样的工序酿成酒，称为米糊酒。其味甜香，常用于节日或款待亲友。度数不高，既能豪饮尽性，又不伤身。

水族糯米酒

水族自己酿造糯米酒，从制作酒曲到蒸米、拌药、发酵、熬酒，全部按传统土法酿制。这种米酒，度数不高，味道醇香，所以人人能喝。水族人有酿诞辰酒之俗。诞辰酒要保存多年，直至该人婚嫁、造屋及过世时才分别开封启用。因其酿制特殊，周期较长，故久享盛誉。制法是：用糯米作原料，采集当地草木本植物制酒曲发酵而成，然后用坛密封置于泥地中。这种密封30余年的糯米酒，会结成状若蜂蜜的浆汁体。饮用时按1：10的比例加入山泉水，或再加入少许鲜米酒即成佳酿。

怒族咕嘟酒

怒族同胞喜欢饮酒，也擅酿酒。贡山怒族的咕嘟酒最有特色。"咕嘟酒"用"咕嘟饭"（用玉米面和荞麦面制成，似年糕）酿制。酿制方法与其他少数民族的水酒酿制法相同，将煮熟的咕嘟饭晾凉，拌上酒曲装入竹篾箩里捂好，发酵三五天后，将其装在罐子里，密封十几天便成美酒。咕嘟酒特别之处在于，饮用时要加水、加蜂蜜，滤去渣，饮其汁。咕嘟酒清醇香甜、开胃可口，即可解渴，又有滋补健身之功效。

布依蜂糖糯米酒

布依族以酒宴客、以酒会友、以酒祝寿、以酒言志、以酒助乐的风俗沿袭千年不衰。他们不仅喜爱饮酒，而且也善于酿酒。米是自己种出来的，酒曲是自己上山采来百草根做成的，所以，酿制出的米酒醇香甘美。每当打开酒坛取酒时，香飘满屋。

蜂糖糯米酒是布依族富有民族特色的酒类。其制作方法是：将浸泡过的糯米先蒸熟，晾凉，加入自制的甜酒曲后拌匀，装入酒坛密封3天后，将温度15°C左右的温开水、白酒曲和少许白酒一起倒入坛内拌匀，再密封；4天后，将适量蜂糖倒入坛内，拌匀，再密封；1～2月后，用纱布将酒糟滤出，坛子里就是蜂糖糯米酒。这种酒清香可口，有健胃、润肺、化痰止咳的作用，营养也十分丰富，是款待客人的好酒。

烧酒及酿制方法

烧酒指各种透明无色的蒸馏酒，一般又称白酒，各地还有老白干、烧锅酒、蒸酒等别称。烧酒起源于唐朝，至宋元以后逐渐普及。明代药物学家李时珍对烧酒的制作方法做了这样的描述："其法，用浓酒和糟入甑蒸，令气上，用器取酒滴。凡酸败之酒皆可蒸烧……其清如水，味极浓烈，盖酒露也。"明代景泰年间郑颙所纂的《云南图经志书·卷之六》曾记载："峨昌蛮……种秫为酒，歌舞而饮，以糟粕为饼，晒之以待乏。"这里所说的峨昌蛮，即今滇西保山、腾冲一带的阿昌族；秫是高粱。种秫为酒，表明这种作物几乎是专用于酿酒的，而"以糟粕为饼"的记述，说明当时当地所酿的酒已不是"连糟而啜之"的水酒，而是蒸馏酒。可以认为，在明代中后期，偏僻山区的少数民族也已经掌握蒸馏酒的技术。

彝族小锅酒

云南哀牢山的彝族善于酿制烧酒，因制作过程中蒸烤是中心环节，故称酿制烧酒的过程为烤酒，又因蒸烤是在家庭小作坊中以小灶、小锅来完成，其成品酒也称小锅酒。小锅酒的主要原料是大麦、玉米、荞麦、稻谷、粟等。酿造小锅酒的过程分两个阶段：一是捂酒饭。将备好的原料粮浸泡透心后煮熟，摊开晾凉，撒上酒曲并搅拌均匀，然后装入瓦罐，封盖发酵。二是烤酒，即用酒甑蒸捂好的酒饭。酒饭内的酒气蒸出后经冷却器凝聚为香醇的酒液。彝家小锅酒醇香爽口，清心提神，是馈赠亲友的佳品。

哈尼焖锅酒

云南红河两岸的哈尼族自酿自饮的烧酒叫"焖锅酒"。哈尼人的焖锅酒具有悠久的酿造历史，哈尼族古歌唱道：

八月的规矩是哪样？

八月的规矩是活夕扎（新谷节）。

看新煮米饭的热气，

绕着家家的蘑菇房；

焖新谷酒的香气，

像云彩一样飘到天上。[①]

焖锅酒的酿造原料以玉米、高粱、稻谷、荞麦为佳，酿造程序上有独到之处。先把选择好的原料粮用清水浸湿，蒸熟后晾凉，撒上酒曲，搅拌均匀，装进一个专用于贮存酒饭的大篾篓里，用稻草把篾篓团团捂紧使酒饭发酵。到酒饭发酵流出汁液时，将酒饭移入瓦缸中，用草木灰和成稀泥封严缸口，再发酵10～15日后，就可以取出焖酒了。焖酒用的木甑是圆台形的，甑内安放一个接酒的器皿。甑的上口放置一个盛冷水的铁锅，锅内的水随时更换以保持冷凉。甑底的水锅里的水加热沸腾后使甑内的酒饭蒸气上升，在甑顶的冷水锅底凝结成酒滴，落入接酒器皿中。因为酒不是按常规在甑子外面接，而是在甑子里面接，故名"焖锅酒"。焖锅酒清澈晶莹，醇厚甘甜，是哈尼山寨节庆必备的饮料。除哈尼族之外，傣族、景颇族、拉祜族等都善于酿制品质极佳的焖锅酒。

怒族、傈僳族蒸酒

云南怒江两岸的怒族和傈僳族称烧酒为蒸酒。蒸酒之名，源自酿造中以蒸为主要工序。蒸酒的首选原料为玉米，也有高粱、稻谷、荞麦、粟。浸泡原粮、蒸熟酒饭、贮存发酵的程序与彝家小锅酒相同，蒸酒时使用的器具则有所不同。怒族、傈僳族所用的甑子是用老树原木挖空而成，甑子的中上部留一小孔插上细竹管，作为出酒槽。锅底加热时，酒气上升遇冷凝聚为酒，落入甑中的接酒器中，再通过出酒槽流出，即为成品蒸酒。

白族烧酒

白族人有"无酒不成礼"之俗。酿制烧酒是白族家庭的重要副业。多用大麦、

---

① 西双版纳傣族自治州民族事务委员会编：哈尼族古歌，云南民族出版社1992年版。

玉米蒸熟后，拌以用48种中草药制成的酒曲，装入陶瓮发酵四五十天，再蒸馏后即成。其中最出名的就是大理州鹤庆县出产的鹤庆乾酒。此酒历史悠久，醇和幽香，饮后回甜，性烈而不易醉人，深受各族人民喜爱。

布依糯米烧窖酒

糯米烧窖酒是布依族特有的一种美酒。其酿法是，首先制作一缸或数缸甜白酒，然后将酒液舀出装于坛中；将坛身放入事先挖好的比坛子稍大的穴洞；坛口放置一个小蒸钵，蒸钵内装清水，并密封坛口；再用谷糠或锯末把坛子周围空隙填满，点燃谷糠让烟火缓慢地熏燃。连续半个月后，坛子里的酒减少三分之二，剩余的酒即是糯米烧窖酒了。一般情况下，五十斤糯米能产七八斤烧窖酒。制成后存储，秋冬取出饮用。糯米烧窖酒其色墨，有丝，醇香味美，既能解渴，又能健胃，是滋阴壮阳、强健身体的佳品。因为制作难度较大，成本高，一般布依族人家只有贵客到来时才取出饮用。

如今，许多生活在云南的少数民族，如彝族、基诺族、阿昌族、哈尼族、怒族、傈僳族、拉祜族等，都能酿造不同风味、不同品质的烧酒。普米族歌谣对蒸烤器具和酿造程序作了极为形象的描述：

朝山头看去，

像有一个大海；

朝山脚望去，

像有一只公鸡在跳；

半山有个小嘴，

醇酒从嘴里流淌。

勤劳的姑娘，

镇守灶台非把酒坛装满；

贪婪的小伙，

寻机舔吃流淌的酒滴，

守在灶旁不离开。

酒坛装满了，

姑娘真高兴，

倒上一大碗让小伙喝，

小伙醉倒三天三夜。[①]

在这首反映酿酒劳动的歌谣中，普米族把蒸烤烧酒的灶、锅、甑的摆放方式喻为一座山，山顶的冷却锅被喻为大海，水锅底火塘中的熊熊火焰是跳舞的红公鸡，木甑上的出酒管则是一张流淌出美酒的嘴。短短的歌谣概括了少数民族烧酒用具的共同特点。

配制酒

种类丰富的配制酒是各少数民族民间医药的重要构成部分。他们利用酒能"行药势、驻容颜、缓衰老"的特性，以药入酒，以酒引药，治病延年。明代初期药物学家兰茂（云南嵩明人），吸取各少数民族丰富的医药文化经验，编撰了独具地方特色和民族特色的药物学专著《滇南本草》。在这部比李时珍的《本草纲目》还早140多年的鸿篇巨著中，兰茂深入探讨了以酒行药的有关原则和方法，记载了大量配制药酒的秘方。

少数民族的配制酒五花八门，丰富多样。有用药物根块配制的，如滇西天麻酒、哀牢山区的茯苓酒、滇南三七酒、滇西北虫草酒等；有用植物果实配制的，如木瓜酒、桑葚酒、梅子酒、橄榄酒等；有以植物杆茎入酒的，如绞股兰酒、寄生草酒、龙胆草酒；有以动物的骨、胆、卵等入酒的，如虎骨酒、熊胆酒、鸡蛋酒、乌鸡白凤酒；有以矿物入酒的，如麦饭石酒。按功效分，少数民族的配制酒有保健型配制酒和药用型配制酒两大类，其中，保健配制酒种类多，用途广，占配制酒的绝大部分。

杨林肥酒

享誉海内外的杨林肥酒属传统配制酒，以其产地而得名。杨林镇地处云南省中部的嵩明县杨林湖畔，早在明初就有繁荣的工商业，酿酒业尤为发达。每年秋收结束，杨林湖畔，玉龙河边，百家立灶，千村酿酒，呈现出一派兴盛景象。传统的酿酒技艺和丰富的药物学知识是杨林肥酒成功的坚实基础。清末，杨林肥酒创始人陈鼎设"裕宝号"酿酒作坊，借鉴兰茂《滇南本草》中酿造水酒的十八方工艺，采用自酿的纯粮小曲酒为酒基，浸泡党参、拐枣、陈皮、桂圆肉、大枣等10余种中药材，同时加入适量的蜂蜜、蔗糖、豌豆尖、青竹叶，精心配制。通过长期的摸索实

---

① 普米族民间文学集成编委会：普米族歌谣集成，中国民间文艺出版社1990年版。

践，于清光绪六年（1880年），酿成了色泽碧绿如玉、清亮透明、药香和酒香浑然一体的配制酒。这种酒醇香绵甜，回味隽永，具有健胃滋脾、调和腑脏、活血和健身的功效。因而创始人陈鼎将其命名为"杨林肥酒"。

刺梨酒

苗族和布依族都会炮制刺梨酒。与一般配制酒不同的是，苗族、布依族配制刺梨酒所用酒基不是蒸馏酒，而是经发酵制成的水酒。其制作方法是：先酿糯米酒，再将晒干的刺梨果盛入布袋，放在酒坛内浸泡。下窖3个月后，取出刺梨渣，即成。刺梨酒色泽呈琥珀色，味美醇香，有助消化、健胃、活血等功效。

鸡蛋酒

彝族的鸡蛋酒是一种具有浓郁地方特色和民族特色的保健型配制酒。配制方法是：

（1）备料。40°～45°纯粮烧酒（以自家酿造的小锅酒为佳）、生姜、草果、胡椒、鸡蛋（以彝山放养的乌骨鸡蛋为最佳）、白糖或红糖。

（2）煮酒。先把草果放在火塘灰中烤焦、捣碎，生姜洗净、去皮、捣扁。备好的草果、生姜和白酒同时下锅，温火将酒煮沸后，加糖；糖完全融化后，撤去锅底的火，但保持余热；捞出生姜及草果碎块，将鸡蛋调匀后，呈细线状缓缓注入酒锅内，同时快速搅动酒液，最后撒入胡椒粉即可饮用。

地道的彝家鸡蛋酒现配现饮，上碗时余温不去，香郁扑鼻，鸡蛋如丝如缕，蛋白洁白如丝，蛋黄金灿悦目，饮后清心提神，祛风除湿。节庆佳期，一碗热腾腾的鸡蛋酒烘托出节日的喜庆祥和；嘉宾临门，一碗香喷喷的鸡蛋酒显示出彝族的真挚与热诚。

松苓酒

松苓酒是满族的传统饮料，其制作方法非常独特：在山中寻觅一棵古松，将白酒装在陶制的酒瓮中，埋于其下，逾年后掘取出来。据说，通过这种方法，古松的精液就吸到酒中。松苓酒酒色为琥珀，具有明目、清心的功效。

第三章　文化承载看酒器

"酒好无好杯，好酒难生辉"，是彝族民间谚语。酒具一般都具有民族性，是民族文化中一个重要的组成部分。从原始的陶制酒具到青铜酒具，竹木酒具、漆饰酒具，瓷酒具以及用玉石、金银、象牙、景泰蓝等名贵材料制成的酒具等，我国各民族在酒具的使用上形成了自己独特的民族风格。

第一节　古朴耐用的自然酒器

自然酒具以自然材料制成。有以动物的角、骨、蹄、爪、皮、壳等制成的，也有用植物的大叶子、花、皮、根等制成的。前者如角杯、禽爪杯、畜足杯、皮酒囊、鹦鹉杯等；后者如荷叶盏、酒杯藤、桦皮酒具和用葫芦、椰壳及竹筒等制作的酒具。

角杯是一种用兽角或畜角制成的酒杯，有牛角杯、羊角杯、犀角杯等。角杯至今仍流行于我国北方和西南地区的一些少数民族中，彝族、哈尼族、苗族、佤族群众仍保留着制作使用兽角酒杯的传统习俗。

彝族的角杯常漆饰彩绘或镶嵌玉石，美观精致。角杯多成双成对地使用，在角的尖端常钻孔或留有突结，以便用绳拴系携带。凉山彝族牛角酒杯有一等牦牛角酒杯；二等犏牛角酒杯；三等黄牛或水牛角酒杯；四等羊角酒杯。角用水煮，掏尽角内血肉，表面刮削干净，髹漆彩绘而成。在哀牢山区，彝族的牛角酒、苗族的羊角酒是贵客临门时必敬的一杯酒。由于牛角杯和羊角杯圆口尖底的造型，客人接过酒杯后，不一饮而尽就无法放下酒杯，这真是"人不劝酒杯自劝"。在昆明西山区核桃箐的彝族婚礼中，有用牛角杯或羊角杯行"轰客酒"的习俗。婚礼的第四天午宴时，新郎新娘手持红线拴系的牛角杯或羊角杯，逐席向客人敬酒。客人必须接过杯子一饮而尽，饮完再敬，直至客人不能再饮，不胜酒力者即刻离席而去，许多人因不能把角杯酒一饮而尽或者怕醉后失态，纷纷逃席。

彝族还使用鹰爪酒杯或禽爪杯，其上部为竹、木或皮胎的酒碗，下部再加禽爪或兽足做成。畜足杯即将牛猪等家畜的蹄子除去肢骨、刮净内皮、装入木质杯模，并经阴干修整，以其撑开的蹄为杯足，杯口和杯身依畜足的形状向一方自然倾斜。

流行于我国南方、北方许多少数民族地区的皮酒囊，将揉制过的整羊皮去头和四肢，除留一腿为囊口外，其余孔眼皆扎住即成；也有用木石等做成酒具模型，再将炮制好的牛皮或其他较厚的兽畜皮紧紧绷在模型上，然后取出内模，打磨修整而成，表面还可上漆做成漆皮酒囊。

木碗木杯。藏族、蒙古族、彝族、门巴族、哈尼族、怒族、傈僳族、独龙族、景颇族、基诺族、阿昌族等都有制作和使用木制酒具的习俗。一般选择树龄较长、木纹细腻、木质坚硬的核桃木、冬瓜木、楸木、椿木和各种栗木。根据各自的文化习惯和所需酒具的容量，截取原木，放置在阴凉干燥处，晾干至透心后，去皮挖空，再削制修整、打磨光滑即可。有的木碗木杯是用树根挖制而成，有条件者再漆上土制朱漆，光鉴悦目，注酒入碗入杯，清冽的酒液在碗或杯子底色的衬映下呈现亮丽的琥珀色。

彝族的酒杯酒壶多用优质的红椿木制成。杯多呈半圆形，上下相连。壶多用木料制作，有一种呈扁形和圆形的壶，盛酒时从底部用细竹管引入，然后再用竹管从上面插入壶内吸食。在西南大小凉山的彝族聚居区，曾有使用"上下酒杯"的习俗。木酒杯为上下相连、大小完全一样的两只酒杯，下杯完全由黑漆涂刷，上杯则在黑漆上又施以彩绘。

竹筒竹杯。在竹海中成长与发展的少数民族与竹有着不解之缘，竹筒盛美酒、作杯饮好酒是硌巴族、傣族、景颇族、阿昌族、彝族、傈僳族、怒族、独龙族等民族重要的生活内容之一。

竹酒筒是一种集盛具饮具为一体的酒具。以竹筒盛酒，酒中渗透了丝丝缕缕的竹子的芬芳清香，别有一番风味。最普通的单节竹酒筒，选择竹林中已成材的竹子，根据容量大小的需求截取其中一节，保持两端的竹节完好无损。锯为两段，一段较长，作为竹酒壶；一段较短，稍微挖1厘米左右长的内壁，酒壶口的外壁也凿去1厘米左右长的外壁作为酒壶的盖子。饮用时，拧开筒盖反置为杯，倒入美酒即可开怀畅饮。

滇西傈僳族、怒族使用的竹节酒壶要复杂得多。竹节酒壶取材于山野中自然弯曲的竹子。一般是截取三节为一段，较粗的一端为壶，保留底部竹节完整无损；上端的竹节凿孔，以装酒和倒酒；较细的一端用刀削成斜口，是为酒壶的壶嘴；中间弯曲的竹节，把两侧壁挖去，唯剩顶部作为酒壶的把柄。这种竹节酒壶浑然天成，美观大方，经久耐用。

竹酒杯有筒制和根制两种。筒制竹酒杯多取山中坚硬的金竹制作，取一节竹筒截成两段，竹节为底，每节竹筒可做两只酒杯。独龙族使用的竹筒酒杯独具特色，他们在竹杯之外缠以藤条炮制的双耳，称双耳竹节酒杯，几乎家家都有。每次饮酒都是两人持同一双耳杯的柄，脸贴脸地同时张嘴去喝，称为"饮同心酒"。滇南滇西的许多世居民族如傈僳族、怒族、独龙族都能制作和使用竹根酒杯。他们将埋在土下的竹根挖出晾干，把侧根须根削去，再把主根挖空为容器，将外部打磨光亮即可。这种竹根酒杯杯壁上留下若干呈同心圆状的疤痕，一连串的同心圆错落有致地排列在酒杯外壁，使整只酒杯看上去拙朴可爱。

许多民族都有庭院种植葫芦的传统习俗，葫芦成熟后掏空籽瓤，在细茎处系上绸带或绳索，外出耕作打猎，走亲访友时用于盛酒。由于外形美观，体积小而容量大，携带使用方便，彝族、苗族、傈僳族、傣族、景颇族等群众都喜欢制作和使用酒葫芦。拉祜族苦聪人不但用葫芦盛装酒，也用剖开的葫芦敬酒，他们传统的吃酒歌唱道："甜甜的美酒，竹碗里倒三碗；香香的酒，木碗里倒三碗；醉人的酒，葫芦瓢里倒三瓢。好酒敬亲人，好肉请好客"。

白族的主要聚居地滇西大理，点苍山所产的大理石又称点苍石，举世闻名。点苍石因石质温润细腻，给人以清新凉爽的感觉，古人称之为"寒水石"。西川节度使李德裕因醉后见到大理石顿觉清醒，故又称为醒酒石。以大理石制作的酒壶、酒碗、酒杯是少数民族酒文化的又一奇观。彩色大理石酒具，丹霞暮霭，山色树影，诱人遐思；水墨大理石酒具，远山近水，虚实相间，清新宜人；而以洁白的大理石制作的酒杯——苍山雪玉杯，则白润如玉，光彩照人。

第二节　古老素雅的陶制酒器

火的使用不仅使人类学会了熟食，而且促使人类进一步学会了制陶，主要陶制酒具有盉（he）、斝（jia）、壶和杯等。这些酒具早期几乎都是食饮兼用或饮酒、饮水兼用，还有用作温酒与调酒的。

我国少数民族在古代就有制作陶酒具。武鸣区壮族先民柳江人遗址出土的春秋战国时期的夔（kui）纹陶罄。佤族的陶酒坛是在饮酒时，中间插一根打通了竹节的弓形竹管，竹管另一端接竹酒筒，利用虹吸管的原理，使酒坛内的酒流出来，供人饮用。

明清以来，铸陶业日趋发展，许多陶窑都以精致美观、实用大方的陶器而远近闻名，如滇西大理赵州的瓦罐、丽江坝和永胜县的土坛、昆明灰土窑的瓦罐都是上好的贮酒器。饮用酒具中，滇南建水五色彩陶"制作精巧，无物不备"，"声如罄、明如水、亮如镜"的建水陶制酒具也成为酒具的上品。

藏族的酿酒器具"朋咱"为陶器，相传由唐朝文成公主传进西藏。此物高约60厘米。甑底用于盛水；中安甑箅，用以装盛青稞；瓶口安放天锅，内注冷水；天锅下设漏斗。粮食久经蒸煮，经漏斗而蒸馏为酒，很是独特。

第三节 高雅富贵的金属酒器

金银酒具非普通人家所能拥有，多为少数民族贵族和土司头人所用，是一种身份等级的标志，大多用于祭祀、大型宴会等活动。蒙古族、彝族、纳西族等民族举行祭天大典时，用金银酒具表示富有虔诚。普通群众也偶有珍藏金银酒器者，但只有在祭祖先等重大活中才使用。滇西纳西族摩梭人祭祖时，对祖先吟唱道："银碗斟满酒，请你睁眼看一看，请你启口尝一尝，喝一口会治好的脚痛；金碗斟满茶，请你睁眼瞧一瞧，请你启口尝一尝，喝一口会治好你的痛。"

少数民族的青铜文化较为发达，因而铜制酒具有悠久的历史，出土文物中已发现为数不少的青铜酒具。明清以来民间已普遍使用铜制酒具、茶具。其中以云南东川的斑铜、红铜酒具最具特色，实用美观。

云南红河哈尼族彝族自治州的个旧是著名锡都，锡制工艺品已有上千年的历史。晚清以来，云南锡制酒具已走上宴席，走进了寻常百姓家。锡酒具晶莹典雅，造型优美，防潮保温，耐酸抗碱。

第四节 鲜亮典雅的漆制酒器

漆木酒具在我国的起源发展，最早可追溯到约7000年前的河姆渡时期。那时我国已有了漆碗，但那时的漆碗还不是专用的酒具，可能主要是食具，仅兼作酒具而已。夏商周三代，髹漆工艺虽得到了迅速发展，但那时主要使用青铜酒具。到了春

彝族漆酒器

秋战国时期，漆木酒具有了较大发展，如楚墓出土的漆勺、漆酒具盒等。但漆木酒具真正取代青铜酒具还是从秦代才开始的，其盛行则是汉魏和两晋。漆木酒具大多外涂黑漆、内涂朱漆，上用朱漆绘制花纹，显得庄重典雅、美观大方。漆木酒具还有防潮、防腐、易清洗和质量轻等特点。今天彝族的酒具也多有各种漆饰彩绘的，如角杯和木杯。凉山彝族的酒具多为木质，以紫红色的漆为底色，绘有黑黄色图案，十分雅致，独具风格。有一种酒壶彝语叫"撒力保"（或"撒那宝"）装有吸管，用木旋制而成；呈圆形扁状如鼓，肩部有吸管；壶肚中心有一空管，酒从底部注入，通过空管流入壶腹，饮酒时从吸管吸饮。

第四章　源远流长酒礼仪

中国自古为"礼仪之邦"，礼在中国社会文化生活之中占有十分重要的地位。人的一生中有降诞、满月、百天、周岁、成年以及婚姻、丧葬礼等。在云南少数民族中，礼与酒也是联系在一起的。

第一节　酒与诞生礼仪

生儿育女，是一个家庭甚至家族中的大事和喜事，一般都要举行一定的仪式，以示庆贺，并通知家族及邻居好友，酒成为必备之物。

彝族人怀孕称为"有喜"，生育称为"喜事降临"。孩子出生后，姑爷到岳父岳母家去报喜，带一瓶酒和一只鸡。生男孩抱母鸡，生女孩抱公鸡。岳父岳母收下母鸡，换只公鸡给姑爷；岳父岳母收下公鸡就换只母鸡给姑爷。姑爷把换的鸡抱回来养，不能杀吃。报喜后，岳母准备煮好的白酒一坛，婴儿背单一个，布一匹，衣服一套，数百个鸡蛋。妇女生育这一月亲戚朋友家都要来送鸡蛋，孩子满月时就要请送鸡蛋的人来吃喝一天。

孩子在满月喜酒时请毕摩来念经，请长辈来剃头，如果留长毛就等孩子长到三岁时才剃。云贵相接的彝寨，孩子满月时有"挂红"的习俗，亲朋扯布去挂在生孩子那家的门上面，门上面钉上若干钉子，布从上面挂了拖到地下。如果新生儿在一岁之前经常啼哭不安时，要抱去路上撞名。孩子的父母准备一瓶酒，一只煮熟的鸡，一锅饭，做一个小木桥，带着这些东西放在山路过小沟的地方，将小木桥搭上，孩子的父母抱着孩子躲在附近观察。看见如果有二十多岁男人从他们做的小桥上过就跑出来拉住，挣下一颗纽扣，抱孩子拜上，要求赐个名字。过桥人起孩子名时，把孩子抱过去，向东南西北四方各拜三拜成为干亲家，过桥人是孩子的干阿爸。双方就在原地烧火热鸡热饭，举杯饮酒，临别时互相告诉名字住址，以后常来常往。

　　三年剃长毛时，孩子的父母要请一次客。要请一长老剃头，剃头前毕摩要念"长毛经"：今天是什么日子？是吉日良辰。今天请我毕摩来，今天我就来念经。念的什么经？念的剃毛经。孩子阿爸听，孩子阿妈听，孩子竖耳听，亲朋仔细听。孩子你知道，你是阿爸骨，你是阿妈肉，骨肉长成体，满月定下来，给你留长毛，把你当小儿。转眼三月满，三月零十天，足足是百天，百天你会笑；转眼六月满，六月零六天，六月你长牙，六月你会笑。转眼日月长，七月你会坐，八月你会爬，九月你会站，站起直打愣。阿妈怕你跌，阿爸伯你跌，眼睛盯得紧，盯你到一岁。你会地上走，你会地上跑。饿了你吃饭，渴了你喝水。你到二岁整，二岁喊阿妈，二岁喊阿爸，二岁会说话。阿爸叫宝宝，叫你拿烟锅；阿妈叫宝贝，叫你拾线团。现在你三岁，三岁剃长毛，弃掉奶熏气，学着懂礼貌。待到七周岁，送你进学堂，接受先生教。从小看到大，三岁能看老，愿你长成才，好把栋梁做。

　　永宁地区在孩子出生后的第三天要举行拜太阳仪式。太阳一出来，产妇的母亲或姐姐就把一根烧燃的松明丢在院子里，产妇左手抱婴儿，右手持一把镜刀、一根麻秆和一页达巴经书，跨出正房到天井里坐一会，让婴儿沐浴阳光，祈求太阳保护。产妇的母亲煮苏里玛酒、猪膘和各种主食宴请村里的老一辈妇女。

　　独龙族的婴儿出生后第七天（男孩）或第九天（女孩），行命名仪式。父母双方的家庭成员和四邻前往祝贺，由孩子的父亲或族内有名望的长老命名。仪毕主人杀鸡宰猪，晋出米酒款待亲友四邻。

　　苗族有小孩拜寄的风俗，谓之"打保脚"。一般是按小孩的八字，找八字好能保护他们的长辈行拜寄之礼。如孩子的八字属水，则拜寄给八字属土的老人，以土保水。通常是选一吉日请保爷及本家亲族等到家吃团圆酒，席间请亲族中一长者讲明某孩子拜某为保爷。举行仪式时所喝的酒均系亲戚自家带来，以表示祝福。然后小孩到保爷家住三日，保爷送银项圈一只给小孩，谓之保命圈。三日后，小孩回本家用酒肉祭祀祖先，自此以后称自家父母为叔（或伯）婶。

第二节　酒与成年礼仪

　　成年礼是一个人由青少年期过渡到成年期的仪式。古代汉族通称为冠礼，一般男子二十而冠。苗族的一支，古董苗男子的成年礼谓之取老名或取尊名。在结婚次日，聚房族老幼于一堂，举行隆重的命名仪式。族中几位老者先秘密商定一名字，然后在堂屋内设酒席一桌，请姐夫坐上席、猜应取的名字，直至猜对为止。姐夫跨出大门宣布命名时，在大门外等待的人要敬姐夫三碗酒，并规定谁要是喊错了老名，罚360斤酒、360斤肉！宣布完毕又敬他三碗酒，说一些祝贺的话。

彝族的剃头仪式也是一种成年礼，须饮酒。彝族儿童从小都要留长毛，长到七八岁，父母就要根据孩子的出生日推算定出剃头的日子。长毛要由孩子的舅舅来剃，已嫁和已定亲的姐妹及姑妈家，都要准备礼物前来吃酒。

第三节　酒与婚嫁礼仪

各民族整个婚姻缔结过程中的每一步骤，从男女青年交友与恋爱、求婚、订婚、送聘礼，直到迎亲与举行婚礼，酒都在发挥着重要的作用。

择偶：窈窕淑女，君子好逑

少数民族青年男女择偶一般均有较大的自由，节日饮酒歌舞常常是男女择偶的最佳机会，如彝族火把节、苗族花山节、白族三月街和绕山灵等传统节日或活动。有的民族有专供青年男女欢聚约会的场所，如傣族和哈尼族的"公房"，楚雄地区苗族的"姑娘房"。各地青年男女约会的时间一般是秋收秋种结束后的农闲时光，地点一般选在山野间，多选择在喝酒吃糖、斗嘴闹笑，或欢快的对歌跳舞中观察和选择意中人。

楚雄山区和禄劝一带彝族就有"吃山酒"的习俗。新春佳节，农闲之余，彝家少男少女携带糖果美酒，欢呼嬉闹着走向美丽幽静的山林旷野间，燃一堆篝火，少男少女以对歌的形式拉开了相互认识了解的序幕。双方即兴对歌作答，逐渐靠拢篝火，含着糖果，饮着美酒，相互逗闹说笑，或趁着酒兴，踩着欢快的竹笛音节，跳起迭脚舞。酒在表情达意中具有十分重要的"催化"作用，故有"三分酒量七分胆量"之说。农历正月十五，云南大姚、永仁县一带的彝族，家家杀鸡宰羊，饮酒对歌，姑娘们穿上自己精心刺绣的服装欢度一年度的赛装节。

求婚与订婚：签订幸福契约

林间吃山酒，田陌共唱和，男女结同心，这毕竟只是婚姻的初始阶段；其后的求婚和定亲，才进入婚姻正式过程。

基诺山区的青年男女相爱后，男方即可在女方家过夜，但暮往早归，这种婚姻的前期阶段可持续1～3年，然后才订婚结婚。巴夺等村订婚分为"衣类""阿下阿尼""阿居瓢"三个阶段。第一阶段，男方请双方三五亲戚及村寨头人连续饮酒三个晚上；其后一两个月进入第二阶段，饮酒一个晚上，主要是给男女双方算八字；第三次是正式订婚，男方带少许钱币和二三十竹酒筒来，请女方送给舅舅，双方饮酒一晚上，订婚仪式基本就结束了。

滇西梁河县阿昌族的求婚订婚也颇富特色，在送过求亲酒、定亲酒及聘礼酒后，新郎一方还要向女方送抱蛋酒。男方家挑选6只新鲜鸡蛋，备好米酒，由媒人带到女方家。当着女方父母亲友的面将鸡蛋逐次打开，放进一个土陶的钵盂内，再加入适量的米酒搅拌均匀后，按长幼尊卑，依次敬给在座的人品尝。

云南香格里拉县三坝一带的纳西族，说亲的媒人是男方的舅舅，前往女方家时随身仅藏有一罐酒和一块糖。舅舅来到姑娘家，开门见山地说明来意，然后自己动手打水洗手，从神柜上取下香炉，烧香磕头。礼毕，姑娘的父母才说："天上成婚时，舅舅牵来白马配金鞍；地上结亲时，舅舅牵来黑牛配木圈。今日阿舅来，不知带的是什么？媒人回答：我只带来甜蜜的红糖，喷香的米酒，它是男方的一颗心，会让你家的姑娘，得到温暖和幸福"！问答结束后，女方的父母若同意求婚，就留下媒人招待茶饭，媒人取出藏在身上的米酒交给姑娘的父母。

佤族在订婚后男方要向女方家行"都帕"礼，送三次酒。第一次送"百来惹"，即氏族酒，为6瓶烧酒和芭蕉、茶叶等，请女方父亲氏族的男性掌家人共饮；第二次送"百来孟"，即邻居酒，为6瓶烧酒，请村寨邻居前来喝酒成为亲事的见证人；第三次送"百来报西歪"，即开门酒，一瓶烧酒，由姑娘的母亲留在枕边，晚间悄悄地饮用，这是为女儿向神灵祈祷，祝她未来幸福美满。

考察各民族的求婚订婚习俗可以发现，酒几乎是必不可少的礼物，而且都是自己家庭酿制的"小锅酒""焖锅酒"。

迎亲：饮酒对歌

婚礼的基本程序由男方娶亲、女方送亲、双方成亲的三部曲构成。虽然在仪式上各地区略有不同，但"酒"是无处不在、无处不有。

迎亲仪式往往在女方家和男方家都要举行。彝族接亲队伍到女方家，由送亲婆把婆家送来的衣服、耳镯给新娘穿戴好后，就吃"递堂酒"。递堂酒是一种礼仪饮酒，就是女方家把亲友请来坐在堂屋中，由送亲先生、送亲婆、新娘三人在堂屋里从右到左转三圈后依次给大家敬酒。摆夜筵，是在女方家进行的婚礼中最热闹的场面，也是气氛达到最高潮的阶段。其形式为：在堂屋内摆两张桌子，坐12个人。桌上放14个杯子，三瓶酒、两钵饭、两把勺子。送亲先生将12个杯子反扣于桌面，将剩下的两个杯子扣在一瓶酒上放于桌下。接亲先生又将桌上的杯子摆正，斟上酒，然后开始说书。

彝族接亲队伍第二天回到男方家，亲友们欢天喜地前来恭贺，鞭炮喧天，热闹非常，欢乐的气氛也达到高潮。迎亲的第二天午饭后，所有亲戚朋友集中在一起喝客人敬贺的甜酒，唱祝酒歌。本家族的男人们则每家提一瓶酒，前往送亲队的住所，饮酒共欢。新娘入洞房后，在里面铺床的同时，外面帮忙的弟兄们将一张大板凳放在大门口，上面摆两坛咂酒，坛内插4～8根（三代人送亲要12根）咂杆，咂杆上扎有花。人们用手握定咂杆，只听酒官一声"请"，咂酒者立即开始吸酒。酒官一声"停"，大家立即停止吸酒。如果喊停未停或者没有咂完规定数量者，要受惩罚，再喝酒。

水族把双方对婚姻责任的承担，放在饮酒之中来表达，叫作吃大酒。接亲时，男方一般去两男两女，其中必有一男一女是结过婚当过家的。他们既代表男方去迎接新娘，又是证婚人。吃大酒时，堂屋中间摆一张长条桌，猪头放在桌子中间，再在两边各放三只海碗，其中一只盛满酒，两只盛半碗酒，下面压着人民币。双方各选德高望重、善于调解纠纷之长者各三人入席。喝上几杯陈年好酒后，双方吃大酒者便开始问答。

问答结束后，两人将"大酒"一饮而尽。接着女方家一人陪男方家一人喝下第一个半碗酒，意即：今后我女方家到你家，如有人讲她闲话，我要找你负责调解平息。女方家又出一个人陪男方家一人喝下第二个半碗酒，意即：我家送女由你负责接到家。这时全场欢声雷动，大家同时喝表示双方亲戚永远友好的"通杯酒"，然后共进晚餐，饮酒对歌，以示祝贺。

婚礼：新人交杯　亲朋痛饮

婚礼是社会约定的婚姻程序中最为重要的环节，各民族在婚礼中有着各具特色的饮酒习俗和规则。

结婚在哈尼人看来是头等大事，他们会把自己的亲戚和朋友都请来喝喜酒。婚礼上，新郎提着酒壶，新娘端着酒杯，依次给客人们叩拜敬酒。客人接过喜酒后说一些祝福语，然后把自己准备好的赐银放到新人的盘子中，最后再干杯。婚礼上新娘要吃下新郎递过来的半生饭，表示自己已经跟新郎结婚了，把半生的饭当熟饭吃，表明自己是心甘情愿的，并且不变心。而新娘也要把自己从娘家带来的圆粑粑给新郎吃，表示自己愿意与新郎团圆，永不分离。

普米族婚礼中最有特色的习俗是"取锁匙"和"锁媒人"。迎新娘时，媒人送礼到姑娘家。新娘上路后，媒人和女方选派出的一个歌手被锁在一间屋里，另有一个姑娘守着门。媒人歌手在屋内对唱，若媒人赢了，守门的姑娘才打开门放行；若输了，就必须喝全寨每家人的一口酒，常有不胜酒量的媒人被新娘的小伙伴们簇拥着、嘲笑着四脚朝天地抬回去。毫无疑问，普米族媒人必须有一张既能唱好曲又能喝好酒的嘴。姑娘出嫁时由若干人组成送亲队伍，新娘陪嫁箱柜的钥匙，由送亲人中的一位年长者保管。到了新郎家，男方要行"取钥匙"之礼才能打开箱柜。新郎家指派一人代表端着托盘，盘内象征性地放少许钱币和一瓶酒，来到送亲人面前，毕恭毕敬肃立，以唱歌的方式讨要钥匙。送亲人回唱后，收下酒和钱，才把钥匙放在盘内。

景东黑彝，新郎新娘双双进洞房"抢床头"，谁先抢到床头坐下谁聪明。坐定后举行"喷酒"仪式。由与新郎关系密切且辈分相同的人斟酒2杯，分别递给新郎新娘。一对新人交腕干杯，口留余酒，相互迎面喷出，谁先喷给对方谁为胜。"喷酒"之后又有"跌架"仪式，新郎新娘再喝一次交杯酒，同出洞房，争着踩门坎，即所谓"跌架"。据说谁先踩到门槛走出洞房，今后的日子中谁就更具有当家做主权。

大理白族婚礼中的交杯酒，由一位夫妻双全、儿孙满堂的妇女主持，事先备好两个完全相同的酒杯，用五色线拴系在一起，新郎新娘坐在喜床上交腕而饮，主持人则祝词"酒要斟斟满，养着儿子做状元；酒要满满斟，养着儿子做先生"等类。哀牢山区的彝族婚礼上，新郎新娘仍用一个葫芦分成的两个瓢盛酒交碗而饮，饮尽后将两瓢合为一个完整的葫芦。据彝族文化研究专家刘尧汉先生考察，这个古礼，象征着新婚夫妇形成了一个合体葫芦，即所谓合体同尊卑，死后灵魂也要进入一个葫芦。

瑶族举行婚礼时要喝"连心酒"。婚礼之夜，男家欢宴宾客，其中首席用5张桌子连成一席，由新郎新娘、媒人、双方父母、哥嫂、弟妹、主要亲朋、族长、外来贵宾入席。新郎新娘给每位宾客斟满酒后，再将杯中酒倒回酒壶混合在一起，再斟到每人的杯中，此酒称"连心酒"。当女方主婚人喊"干杯"后，连心酒宴便开始。新郎、新娘向每一位长辈、亲友敬酒。每敬人一杯，自己便陪饮一杯。

满族婚礼正午要祭祀神灵。院中设一神桌，供上猪肘，三盅酒放在碟子内，另放尖刀一把。新郎新娘面向南方跪在神桌前，萨满单腿跪在神桌左边，唱《阿察布密哥》，祭歌分三段，每唱完一段就用尖刀割一片肉抛向空中，然后端酒一盏，举至齐眉处，再泼到地上。歌完，鼓乐齐鸣，将婚礼推向高潮，这叫"撒盏"。晚上要在洞房内吃交杯酒、行合卺礼。洞房内摆一桌子，新郎新娘手挽手绕桌三圈，由全福人斟满两杯酒，夫妻各抿一口，然后交换酒杯，再饮一口，是为合卺礼。接着吃子孙饽饽和长寿面，然后双方争坐被上，以为吉兆。新娘新郎一改白天的拘谨情态，互相争着坐被上，难分高低上下，结果常是二人同时坐在被子上面。新婚之夜，洞房蜡烛，彻夜不熄，外面有人唱喜歌，叫"响房"。有人将黄豆撒向洞房，祝愿一对新人婚后富足有余、早生贵子。

第四节　酒与丧葬礼仪

丧葬礼也是人生礼仪之一，在丧葬礼的各个环节酒都有一定的作用。

携酒报丧

人死后要向亲友通报死亡的消息，许多民族都有携酒报丧的习俗。红河两岸的布依族在人死后，鸣火枪三响并击鼓向寨中报丧。随后死者的子女或侄子到亲友门前跪拜，并呈上带去的一壶（瓶）酒和一张白布孝帕。楚雄县哨区彝族赶尸亲者，帽下齐眉夹一小片白麻布，带一只鸡一壶酒前往后家，到了堂屋门前即止步，将鸡抛进堂屋，酒置于门前，然后在门外跪拜。后家捉鸡收酒准备奔丧事宜，整个报丧过程甚至不说一句话。各民族普遍认为报丧会带米污秽的邪气，待报丧人走后，被通知的人家要抿一口酒在屋内喷酒，并用扫帚作扫除状，以期驱除邪气。有条件的人家在报丧人离开后还要请祭司念经驱邪。

以酒奔丧

云南路南圭山地区的撒尼人，出殡之日全寨男女老少各自携带酒菜参加葬礼，男人在棺前，女人在棺后，直达墓地，棺木入土后，饮酒吃菜，劝慰丧失亲人的家属，最后才相携相依回寨。

奔丧所携带的物品中，最重要的就是酒。酒的数量多少不等，依据奔丧人与丧主的亲疏关系而定，多至数十斤，少至一二斤。云南武定、禄劝彝族聚居区奔丧的远亲带一斤酒、一只鸡，至于近亲则须牵羊背酒前往；滇西宁蒗彝族自治县境内的摩梭人，亲友奔丧只需拿一瓶酒、几片茶叶前往即可；子女奔丧则需牵牛赶羊，酒则以罐论；如果死者是女性长者，舅父奔丧携礼最重。奔丧人把酒瓶或酒罐放在棺材前的供桌上祭奠后，由主丧者收下，送葬结束后，纳西族摩梭人有"回礼"之俗，牵羊者回一条羊腿，牵牛者回一条牛腿和牛头，携酒者瓶底或罐底留少许酒，并声明礼物是死者回送的，由奔丧人带回家祭祖。

美酒祭祀祖先灵魂

许多民族认为人死而灵魂尚存，因此在尸体入土垒坟后，还有一系列善后活动。彝族普遍认为人有三魂，死后一魂回到祖先古代居住地；一魂留守坟地，保佑子孙发达；一魂在家接受供奉，保佑家人顺利吉昌。后两魂若不认真供奉就会成为野鬼，危害生者。死者下葬后，还有收魂、做祖灵牌、作法念咒以超度死者等活动，属丧葬活动的延续。

普米族视死如生，停尸举丧期间，死者灵前的美酒不断，送棺入土前，仍要敬献好酒，使死者载酒西归。滇西宁蒗彝族自治县托甸一带的普米族挽歌（吊哦调），唱道：

在你临走之前，我把美酒给你敬献，这碗香醇的美酒哟，来自辛勤的汗水。这碗醉人的美酒哟，是普米的四个家族敬献。

老大坐在上方，他是藏族，我向他找来青稞。

老二坐在中间，他是纳西族摩梭人，我向他找来苦荞。

老三坐在下方，他是普米族，我向他找来大麦。

老四坐在席尾，他是汉族，我向他找来谷米。

高山的牧羊人，送来了苦酒曲。江边的放猪人，送来了甜酒曲。村里的放羊人，送来了酸酒曲。苦甜酸酒曲制成了酒哟，酿就的苏里玛最香甜。

今天呵，我把苏里玛装在金角银角里，双手捧放在你面前。你用眼睛看一眼，眼睛会更明亮；你用舌头尝一尝，说话会更快；你用手拿一拿，做什么事都成功；你在脚上滴一滴，可以走到任何地方。

你的头会变成海螺树，你的眼睛会变成日月，你的牙齿会变成星云，你的舌头会变成彩虹，你的血液会变成大海，你的骨头会变成高山，你的肠子会变成大路，你的四肢会变成四个方向。你用一只手，拿着敬献的檀香树杯；你用一只手，拿着敬献的酒碗，去寻找你的幸福吧，去寻找你的欢乐！[1]

---

[1] 普米族民间文学集成编委会编：普米族歌谣集成，中国民间文艺出版社1990年版。

七 灵捎美酒

在少数民族群众各种丧葬习俗中，还有一种叫作"寄酒"的丧葬活动。所谓寄酒，即生者请死者的灵魂为自己已经逝去的亲人捎去美酒，这种习俗至今仍完整地保留在滇西的普米族、纳西族摩梭人等群众中。

在风光秀丽的泸沽湖畔，纳西族摩梭人不仅自己喜欢饮酒，而且相信死去的人也最喜欢饮酒。由于生死相隔无法沟通，他们就有了请刚刚去世的人为自己早已去世的亲人和祖先们捎去一份美酒的习俗。纳西族摩梭人称集中奔丧的一天为"布足"。布足之日，除至亲好友纷纷携酒前来奔丧外，四乡八邻不同氏族的纳西族摩梭人也带着酒赶来，供在棺材前面，言明是托死者把酒带上，见到他们的亲人和氏族祖先时，送给他们。寄酒者要双手捧酒碗，高举过头，长跪在死者灵前，由经师唱完寄酒歌，才算完成托付。寄酒歌唱道：

不久前这里路过普米人，想托他把美酒寄给祖先，只因不懂普米话，没有办法把酒寄到。

隔天来了汉族人，想托他把美酒寄给祖先，只因不懂汉族话，没有办法把酒寄到。

一天路过赶马人，赶马人他忘性大，怕他把酒丢路上，不敢把酒寄给他。

一天大雁头上飞，大雁说它忘性大，怕它把酒洒在白云间，不敢把酒寄给他。

一天野鸭飞过村，野鸭说它忘性大，怕它把酒倒在河草滩上，也不敢把酒寄给他。

我们这一村，你将见到我们祖先的面，烦你带信并不重，口信像轻风，带信并不难，风儿会帮你带。[1]

---

[1] 云南省民间文学集成办公室编：云南摩梭人民间文学集成，中国民间文艺出版社1990年版。

第五章　别样风采酒风俗

以酒交友，以酒结谊，以酒叙情。热情好客、以酒待客几乎是少数民族的共同特点。禄劝彝族聚居区的群众，在亲友酒足饭饱离席之际，还要唱留客歌，宣称"米酒多多有，喝了九小坛，还有九十九"。

第一节　热情豪放的待客习俗

酒在人际交往中的重要作用，主要是以酒飨客。酒主要用来表示对客人的欢迎和尊敬，同时也表达和促进主客相聚的欢乐。各民族人民都很重视以酒待客，彝谚说："汉人贵茶，彝人贵酒"。

在苗族地区，牛角酒是一种比较著名的、尊贵的待客酒。因牛角底尖，不可置放，客人必须一饮而尽。苗家自古就把牛当作宝贝，与牛做朋友。母牛生小犊，要给母牛喂豆浆、糯米稀饭；逢年过节，要给牛喂些肉、鱼、蛋，甚至要用竹筒给牛灌点酒，以表彰牛的辛劳，让牛与人一起分享节日的欢乐。耕牛死后，苗家人十分痛心，为了纪念，便锯下牛角制成酒杯，悬挂屋中。客人进主人家门，需饮一牛角的"进门酒"，客人入座则饮三杯入席酒，然后才共饮"转转酒""交杯酒"等。逢年过节或遇喜庆，或有贵客来到，人们便用牛角杯饮酒敬酒，既表示对客人的尊敬与爱戴，也表示对牛的怀念。有的还常在寨门的木楼里挂一对牛角杯，贵客进寨，便由穿古装的寨老或盛装的少女，双手捧着牛角杯，向客人一一敬酒，是苗族社会中一种最尊贵的迎客礼仪。

送客酒也是苗族人民的一种礼节。客人就要回家，主人意犹未尽，临别时还以酒相送。主客各执盛满美酒的酒碗，主人一边唱歌送客，一边和客人痛饮美酒作别。如离寨的是尊贵客人，送客酒则更为热烈动情。人们先在铜鼓坪上踩响铜鼓，跳起芦笙舞，召唤全寨出来为贵客送行。男女老少陆续走下木楼，来到铜鼓坪上，围着铜鼓跳起舞来。铜鼓坪上放一条桌，桌上置酒数碗、红绸几匹。寨老宣布送客，便将红绸斜挂在客人身上，顿时坪上一片欢腾。人们一个接一个地走上前来，唱一支送别歌，敬一口送客酒，再将五颜六色的花带系在红绸上。客人步出寨门时，主人再劝客人喝一口才最后分别。

藏族人用青稞酒招待客人时，先在酒杯中倒满酒，端到客人面前。客人要用双手接过酒杯，然后一手拿杯，另一手的中指和拇指伸进杯子轻蘸一下，朝天一弹，意思是敬天神；接下来再来第二下、第三下，分别敬地敬佛。在喝酒时，藏族人民的约定风俗是：先喝一口，主人马上倒酒斟满杯子；再喝第二口，再斟满；接着喝第三口，然后再斟满。往后就得把满杯酒一口喝干了，这样做主人才觉得客人看得起他。客人喝得越多，主人就越高兴，这说明主人的酒酿得好。藏族敬酒时，对男客用大杯或大碗，敬女客则用小杯或小碗。藏族善歌舞，饮酒时少不了唱酒歌、跳锅庄舞。

蒙古族的待客礼仪很多，归纳起来，大体有敬烟、敬酒、献哈达、献德吉、献整羊整牛等。主人敬酒，客人要双手接过，不可用左手给主人递东西。酒宴上，不论受酒或敬酒，都要把挽起的袖子落下来。斟酒敬客，是蒙古族待客的一种最普遍的传统礼节。蒙古族人认为，美酒是食品之精华、五谷之结晶，拿出美酒敬献客人，可以表达草原牧人对客人的敬重与爱戴。

布依族重礼好客，贵宾到来，必有进门酒、交杯酒、便当酒、转转酒、千杯酒和送客酒等六道酒礼。若是敬献猪肉，是祝客人来年养大猪，收成好；若是敬献鸡肉，鸡头奉给首客，象征吉祥如意，鸡翅奉给次客，表示腾飞，鸡腿奉给三客，意为脚踏实地。宴席中还唱《祝酒歌》和《宵夜歌》，前者是殷殷劝酒，后者要将餐桌上的所有物品食品都一一唱出，表现出他们的心智和才华。

布依人敬酒，都敬三大碗便当酒，人称"三巡酒"。客人每巡都得喝干一碗，若想逃酒，姑娘们就会不停地唱敬酒歌，情真意切，客人很难推却。布依山寨待客，头天由主人家招待，次日起就由全寨轮番宴请，这种吃"百家饭"的习俗体现着布依人热情好客的传统。每当农闲，家家户户都要自酿米酒，自家饮用。此类酒制作方便，度低味醇，甘甜可口。凡婚丧嫁娶、立房、满月、祝寿等，都用便当酒宴请客人。

热情好客的壮家人向宾客敬酒有敬12杯之俗，"12"与壮族传统文化有待解之谜。壮族神话里早期有12个太阳，被郎金射下11个；一年分为12个月，一天分12个时辰；人的属相定为12；活人有12个魂（命）；嫁娶时男方需送女方12块银圆、

12斤酒、12个大糯米粑粑，送亲人数为12人；祭神供12盅酒；送亡人灵魂归祖要涉12条河、过12座桥、登12座山、经过12寨等，故向宾客敬酒有12杯之俗。壮族人敬客人的交杯酒并不用杯，而是用白瓷汤匙。两人从酒碗中各舀一匙，相互交饮。主人这时还会唱起敬酒歌：锡壶装酒白连连，酒到面前你莫嫌，我有真心敬贵客，敬你好比敬神仙；锡壶装酒白瓷杯，酒到面前你莫推；酒虽不好人情酿，你是神仙饮半杯。

哈尼人有喝焖锅酒的礼俗。客人进屋后，先请入座火塘边的上席，随敬一碗喷香的米酒，称为"喝焖锅酒"。哈尼族很注重待客礼节，吃饭时总要摆一桌丰盛的饭菜，请客人到上席就座。先给客人杯里斟满酒，然后再给其他人斟酒。举杯前，主人轻念祝词后，便用食指从自己杯里醮一点酒，分别在自己面前的桌面上和自己的脑门上划一个"一"，表示驱邪，祝福全桌人幸福安康，然后大家举杯畅饮。喝酒时小辈不能陪客，主人不能跷二郎腿。送客时要往客人的口袋里装上一点礼物，如鸡蛋饭团等。如果让客人空手而归，则被视为失礼。哈尼族的人在敬酒时，一般只会敬三杯，因为在他们的眼里，三是非常美好并且吉利的数字，最重要的是哈尼族的人认为三杯酒是最适合的，喝三杯有助于自己的身体健康，喝多了会伤害人的身体。在喝够三杯酒后，他们就不会再敬酒，如果愿意的人可以继续喝。如果不想喝就可以吃饭了，人们绝不会故意把客人灌醉，他们只是希望客人能够高兴和健康。

普米族有句俗话："新坛子倒出来的第一碗青稞酒，要敬给远方来客；刚烧涨的茶罐里盛出的第一杯茶，要端给远族兄弟喝。"

怒族至今仍保留着独特的迎客宴习俗。客人到家后，主人先拿一些玉米粑粑、玉米糕等食品请客人吃。全寨各家各户献出自己珍贵的鹿子肉、兔子肉、松鼠肉等，主人拿回家烧烤后舂碎拌进蒸熟的糯米饭中，装到簸箕里，这就是宴席的主食，接着主人又取出自酿的玉米酒。酒席开始，主人和客人慢慢品尝叙谈，当客人酒足饭饱时，主人便双手捧起盛满酒的竹筒走到客人面前说："你就像天上的星星，从远方到来，在我们心里永不消失，愿我们的友情像滔滔的怒江水长流不断"。随后与客人共捧竹筒，互相搂着脖子，共同张口一饮而尽，这叫"双边倒"，亦称"同心酒"，是宴席的高峰。

第二节　饮酒习俗

在人类社会中，任何一种食物的食用方式都会受到社会结构和文化特征的影响，从而形成丰富多彩的饮食文化和习俗。其中酒的饮用规则和饮用习俗又是最为复杂、最具文化特色的。

火塘酒

火塘酒，即在火塘边饮酒及其相关的规程。火塘是少数民族生活的重要组成部分，在少数民族地区，居家饮酒几乎都离不开火塘。

在传统的彝族社会中，火塘"上方"指背墙面门的位置，这个位置离供桌最近，是家庭中男性长者的专座；纳西族摩梭人则正好相反，火塘上方是当家妇女的座位。以前大多数民族占有火塘上方的，为男性长者，依次是女性长者、长子、次子，儿媳女儿等在火塘边几乎没有位置，有也是在火塘下方，专事添柴加火之职。

居家围坐火塘饮酒，斟酒人一般是家庭的长子，第一杯酒要敬给男性长者，然后是女性长者，平辈者依年龄长幼顺序斟满。若有宾客临门，第一次斟酒要由男性长者亲自执壶，为宾客斟满后，再移交酒壶给长子，由其依次斟满。饮酒时要先敬客人或长辈后才能饮用，饮用时碰杯而不干杯，饮多饮少随意。

傈僳族、怒族、独龙族的火塘酒，更多的是追求一种宽松舒畅、热烈欢快的生活氛围。独龙族的酒歌唱道：我家的猪养肥了，我家的米酒煮好了。欢迎大家一起来，围着火塘吃肉喝酒。以后你家杀猪，大家又到你家去。这是我们的习惯，这是我们的习俗。把火塘里的火添旺，饱饱地吃，多多地喝，喝够了酒才好唱"珍珠"，明天砍柴烧山才有力气。

咂酒

咂酒不是酒，是一种独特的饮酒习俗，是借助竹管、藤管、芦苇秆等管状物把酒从容器中吸入杯、碗中饮用或直接吸入口中饮用。因选用吸管的不同，咂酒又称竹管酒、藤管酒等，流行于四川、云南、贵州、广西等地的彝族、白族、苗族、傈僳族、普米族、佤族、哈尼族、纳西族、布依族、壮族、侗族等民族之中。以咂酒法饮用的酒都是水酒，咂酒有冷咂热咂之分。冷咂即搬出酒坛，将吸管插入坛底吸饮；热咂是把水酒放在锅里加热或者直接把酒坛架在火上，边加热边饮用。吸管都是一插到底，一边饮用，一边加入冷开水，使坛内或锅内的酒液保持在相同的水平，直到酒味全都丧失。

这种饮酒方法在西南各民族中曾长期盛行，是待客的最高礼节。明代旅行家徐霞客游历滇中，在洱海边的铁甲场村民家晚餐时，这种独具特色的饮酒方式令徐霞客大开眼界。滇南元江两岸的傣族社会也有类似的饮用水酒方法。佳节良宵，人们在宽阔的场坝上置盛满水酒的桶或大罐，其间插入竹管若干根，人们环绕着酒坛轻歌曼舞，渴了凑近酒坛对着竹管喝一口，清清喉咙再唱；累了凑近酒桶吸一气，振振精神又跳。贵客临场则欢迎加入歌舞，一曲舞罢，众人簇拥宾客到酒坛前，主持者执管相邀，客人插管，众人才插管入坛，同饮共欢。

普米族、佤族等群众就是用竹管将酒吸出盛在葫芦和碗里饮用。酥理玛调是反映普米族群众酿制酥理玛的过程和享用酥理玛的喜悦之情的传统歌谣。"五月端阳节，上山砍来岩金竹，做成哑杆来哑酒。哑杆像条弯柳枝，引来土坛里的酥理玛。香醇的酥理玛，倒满金边银碗里，酒浓起泡泡，阿公阿爹举碗笑哈哈，喝一口醉了心"。

转转酒

转转酒是若干人围坐，共用一只酒壶或一个酒碗，按一定的方式轮流传递同饮的饮酒形式，流行云南四川等地。漫游西南少数民族地区，经常可以看到田边地头，山前路畔，街道集市，三五人团团围坐，中间是一把酒壶和一个盛酒的土碗，众人按一定的方向传递着酒壶或酒碗，饮一口酒，说一段笑话，直至酒尽壶空，才欢笑着分手，尤其在赶集的日子。

彝族、傈僳族、苗族、怒族等饮转转酒的情况更是司空见惯，流传于哀牢山区的彝族歌谣就唱出了转转酒的实质。

"彝家的祖祖辈辈，自古心胸开阔。我们喜好白酒，我们尊重贵客。不问你来自何方，不管你穷得怎样，只要你走进彝山，我们就是一家；只要你真心实意，我们就是朋友！哪怕穷得穿不上衣，哪怕只有半勺米，哪怕只有半只鸡，哪怕只有半块乔粑粑，哪怕只有一口酒，我们都要一人吃一半，我们都要一人喝一口。因为你是贵客，因为你是彝家的朋友，因为我们是同一个祖先的后代，本来我们就是一棵树上的叶子"。[①]

---

① 云南省民间文学集成办公室编：云南彝族歌谣集成，云南民族出版社 1986 年版。

转转酒的传说，相传有一座大山，上面住着汉藏彝三个民族，结拜为兄弟，汉族为大哥，藏族为二哥，彝族为三弟，逢年过节都团聚在一起。有一年，三弟彝族开荒收获了许多荞麦，磨乔面后煮了很多，请二位兄长前来分享。第一天没有吃完，第二天泛出了浓烈的酒香，舀进碗后，三兄弟你推我让，谁也舍不得喝，从早转到晚也没有喝完。后来神灵告知，只要辛勤劳动，喝完后又会有新的，于是三人才放怀转着喝酒，结果个个喝得酩酊大醉。

转转酒也是苗族人民热情待客的一种方式。一般是逢年过节或招待宾客酒至半酣、热情高涨时进行。主人在每位客人面前斟半碗酒，各人都把自己的酒传递给左边的人，要求每个人都用右手接酒。人们围成一圈，由场中年纪最大的老人先饮，表示尊敬老人。接着每人按顺序饮尽，叫一轮；接着主人再斟第二轮酒，全部喝完后，又起一轮。如此往复不断，直到酒醉方休。

交杯酒和交臂酒

壮族宴席间敬酒方式有交杯酒、交臂酒和转转酒等不同形式。主客敬酒时，客人饮主人杯中的酒，主人饮客人杯中的酒，称交杯酒或串杯酒。主客均右手执酒杯，两臂相挽相交，各饮自己杯中的酒，称交臂酒。主客围桌而坐，相互之间同时敬酒，各人饮其身旁亲友杯中之酒，称转转酒。壮族饮交杯酒时，要杯杯斟满，一饮而尽。

交杯酒也流行于苗族和水族地区。其喝法有三种，前两种与上述交杯酒与交臂酒相似，第三种是互相交换酒杯酒碗而饮。你喝我的酒，我喝你的酒，表示肝胆相照，以心换心的真诚友谊。水族以酒为礼，以酒为贵。喝到兴起时，主人提议大家举起酒杯，这是要喝团圆交杯酒的意思。于是每个人用左手接过别人送来的酒杯，右手拿起送给别人的酒杯。当大家都手拉手地举起杯来时，由贵宾中年岁最大的人先喝，然后按从左到右的先后顺序喝。每个人举杯喝酒时，其余的人都要为他助兴，齐声喊"秀，秀"，即干杯的意思。

过关饮酒

过关饮酒即在迎客的道路上，设置一道道关口，客人必须先饮酒然后才能通过的一种待客饮酒方式，表现出主人以酒迎客的至诚与坚决，流行于我国南方苗、瑶、侗和布依等少数民族中。

瑶人嗜酒好客，瑶族待客有饮"三关酒"之俗。凡喜庆事，贺客到，唢呐喜炮响起，主人家就端酒在屋外组成三道关。每一道关必敬每一位客人两杯酒，称为

"三关迎客六杯酒"。

拦路酒是苗族人民的一种迎客习俗，凡客人进寨，人们便在门前大路上设置拦路酒，对客人唱拦路歌，让客人喝拦路酒。拦路酒的道数多少不等，少则三五道，多至12道，最后一道设在寨门口。客人要一道关一道关地喝完才能进寨门，既表示对客人欢迎之热诚，又表示主人待客之盛情。

拼伙酒

拼伙酒是许多民族共有的饮酒习俗，以参加人共凑份子的形式饮乐；又因饮酒歌舞多在春暖花开时节，地点多在远离村寨的林间草地，参加者大都是花季少男少女，拼伙酒也叫"吃山酒""饮花酒"。云南省大姚县彝族县华山插花节，巍山县巍宝山二月八会，贡山县怒族花山节，滇东南苗族踩花山等节日，均与拼伙酒的饮用习俗有着直接的渊源。

"三月三，耍西山"。五百里滇池畔，碧鸡山"花酒会"和玉案山"饮花酒"源源流长，影响极广，至今仍盛行不衰。玉案山花酒会据传为大理国国王段素兴所创。素兴曾在昆明营建宫室，并筑春登堤、云津堤，广种奇花异草。春暖花开时节，遍邀青年男女携带酒食游玉案山，并开沟引水作九曲，水上浮小木板，木板上置杯盛酒，男女列坐，酒船飘到谁面前停下，就该谁饮酒，并斗草簪花（斗草是古代流行在女孩子中的一种游戏，簪花意思是插花于冠）为乐。

同心酒

两人共用一酒具，或并立、并坐、并跪、并蹲，两人搂肩交颈，耳磨脸贴，一个用左手，一个用右手，同时持杯（筒或碗），嘴凑在一起，同时饮酒，称同心酒。酒可一饮而尽，也可轻抿一口，说唱一段，再饮一口，如此再三，直至兴尽酒尽。使用的酒具有木碗、竹筒、牛角杯、羊角杯、猪蹄杯等。此饮酒方法许多民族均有，以彝族、苗族、傈僳族、怒族、独龙族等群众中最为常见。

傈僳族同心酒体现了傈僳族热情豪放的民族性格。傈僳族世居高山大川，居住分散，一旦相聚在一起时就用酒和歌舞表达双方的深情厚谊，特别是用"三杯酒""双杯倒"等饮酒方式来表达自己的情感。傈僳人家将这种多姿多彩、独特的饮酒方式称为"同心酒"，同心酒展示的是傈僳族对亲朋友人的深情，展示的是"同心同德"。同心酒有六种不同的喝法，寓意也各不相同。

第一种是"亚哈巴知"（石月亮酒）。怒江大峡谷深处的福贡县利沙底乡境内的高黎贡山上有一天然岩石空洞，犹如一轮明月高悬西天，傈僳语称石月亮为"亚哈巴"，它是所有傈僳族人民心中的太阳，是傈僳族追祖寻根的发源地。亚哈巴知体现傈僳族追求团结、尊重朋友、纯洁真诚的品格。饮酒时众人围桌而立，右手端酒杯，同时用左手挽住朋友们或客人，整个场面如同满月。在唱罢祝酒歌后，众人一齐说"一拉秀"（意思是一口干）。

同心酒

第二种是仨尼知（三江并流酒）。傈僳族是金沙江、澜沧江和怒江的主人，主要聚居区就在如今"三江并流"风景区的核心地区。傈僳人视三江并流之水为美酒，把三江并流与饮酒结合，展示了三江的美、人与自然的和谐。喝"仨尼知"时，三人左手搭靠在一起并靠近，右手端杯逆时针方向缠绕形成三江之流，象征着三人携手共创美好明天。

第三种是"燃卡知"（勇士酒）。勇士酒也称英雄酒，是傈僳族勇士"上刀山、下火海"时的饮酒方式，喝过此酒意味着有无比的勇气和战胜一切艰难险阻的决心。一是长辈或尼扒（祭司）送勇士的"壮行酒"，敬酒者同时用手端两杯酒给勇士并说道："尼子知多"，勇士饮毕拱手而谢。二是勇士胜利归来，长辈或尼扒手端两杯酒，饮酒时勇士先将头偏朝右为半蹲式，尼扒将头偏朝左边示意接受，并将左手中的酒敬给勇士喝；勇士将头偏朝左边，尼扒将头偏朝右边示意肯定，并将右手中酒敬给勇士喝，这样勇士就成为凯旋的英雄。

第四种是"普花知"（发财酒）。普花知是傈僳族人民在与自然的斗争中，求顺利和发财的美好愿望。饮酒时两人手端酒杯交叉勾住对方手腕，同时用手扶住对方，下肢也交叉，横看竖看都像一个阿拉伯数字"8"。上下两个"8"字，表达"发了又发"的美好愿望。

第五种是"斯加知"（思念酒）。斯加知是傈僳族同心酒中最常见的方式，也称弟兄酒和兄妹酒。斯加知是对远方来的朋友、客人和亲人表达深情厚谊的方式。饮酒时两人面对面，右手搂对方颈部，左手轻扶对方背脊，先说"尼迟知多"，再喝杯中酒。另一种是两人搂肩脸贴脸，嘴靠拢，同时饮完杯中酒，以喝完一滴不洒为佳。

第六种是日师知（长寿酒）。傈僳族有尊老爱幼的传统，有敬老胜于敬天地之说，敬天举过头，敬地弯腰，低于长辈杯子、碰杯共饮。敬长辈时，晚辈双手捧酒杯半跪三磕，向前敬给长辈老者。

马上敬酒

在长期的社会历史发展中，马、骡是山地民族最重要的代步工具。贵客临门，主人恭候，在来宾没有下马之前，即敬献美酒，是为马上敬酒。马上敬酒是彝族、白族、纳西族等民族接待贵宾的最高礼节，多用于隆重庄严的场合。

如今在有公路通到的山寨，贵宾进村时人们夹道欢迎，来宾尚未下车，主人就会敬上一碗清冽甘甜的美酒。来到少数民族村寨的客人，如果受到马上敬酒或车上敬酒的礼遇，表明山寨已经把你当成了最尊贵的客人，呈送到胸前的美酒，是绝无推辞余地的，一定要双手接酒，举杯齐眉，以示谢意和回敬，再一饮而尽，确实不胜酒力者，也要把酒送到嘴边，多少喝一点才行。

歌舞劝酒

少数民族大多能歌善舞，而将民族歌舞引入酒席劝饮助兴，是少数民族酒文化的又一独特景观。

歌舞劝酒的习俗尤其盛行于彝族社会中。佳期良宵，彝家山寨的宴席是由一个小型的管弦乐队组成敬酒劝酒的队伍，一人执壶，一人捧牛角杯，一人唱曲，一人吹奏葫芦笙，一人吹奏竹笛或长箫（近年多吹奏口弦、口琴），一人弹三弦，边演奏边舞蹈来到尊贵的客人面前。执壶者上酒后退离一边，捧杯者举杯齐眉，宾客不饮，歌舞不止。宾客则应起身肃立，等劝酒一曲唱罢，也以歌致谢，双手接酒齐眉，回礼致谢后，或一饮而尽，或浅尝轻抿，再双手举杯齐眉致谢，还杯敬酒者。

云南峨山彝族自治县小街山区的彝族在节庆佳期或新婚庆典时，客人们在铺着青松毛的场院中席地而坐，饮酒谈天，主人（一般是男性长者）率领一群青春洋溢少男少女款款走来，把客人围在中间，逐次劝酒。主人自持一杯，身边的一个少男执壶上酒，一个少女接酒举杯待立，主人舒缓而歌："敬你一杯酒哟，请喝上一杯杯，喝上一杯酒呀，表表彝家的心意。敬你两杯酒哟，请喝上两杯杯，喝上两杯杯酒哟，祝愿你快乐。敬你三杯酒哟，请喝上三杯杯，喝上三杯酒哟，祝愿你幸福"。

秆秆酒

彝族的传统酒类是秆秆酒。秆秆酒用坛子盛装，饮用时，用细竹秆插入坛中吸饮或者把酒引入酒杯饮用。秆秆酒因用坛子盛装，故以"坛"作为计数单位，现在有以瓶装的一瓶酒也称为一"坛"酒。

喝秆秆酒的又一特色是采用萨玛（刻度、标记）制度。即在一竹片上钻一个小眼，插入一根小竹条，喝酒时，将竹片横放在酒坛口，小竹条朝下，即成萨玛（相当于一杯酒）。

秆秆酒属水酒类，酒度低，一般在20°～30°之间，酒味醇香浓甜，老少皆宜，一年四季都适合饮用。每次喝时，将水倒入坛中与酒混合，水倒至与坛口平。喝酒者须用秆秆喝酒，直至萨玛的小竹条完全露出，作为敬了一个萨玛。再加满水，第二位饮者也须将萨玛小竹条喝得完全露出，如此反复，直至酒味淡如水。一坛大的秆秆酒可以喝好几天，一般过年泡一坛秆秆酒足矣。

喊酒

苗族有一种独特的饮酒方式，扯着耳朵喝喊酒。苗家喝酒不兴猜拳，而以喊酒助兴。饮酒时先各自随意饮，酒过三巡后，主家端起酒杯或酒碗站起来，众人也端酒杯站起来。甲将自己的酒传给乙，乙也将自己的酒传给丙，依次传换，最后一位将自己的酒换给甲。大家站在一起手拉着手，弯腰搭背，围成一大圈。主人或其中一人诵念酒理、唱酒歌，到高潮处大家放开嗓门喊"唷——唷！"灌了一轮酒，又互相喂肉。酒至半酣，便开始扯耳朵饮酒。有两种方式：个体对扯：甲与乙互相换酒，双方右手端酒杯递给对方，同时伸出左手扯住对方的耳朵，同时喝干对方递过来的酒。集体轮扯：大家都站立起来举起酒杯，甲递酒给乙，同时伸手扯住乙的耳朵，乙依样递酒给丙并扯丙的耳朵……都扯着耳朵后，大家同时喝下上面的伙伴递来的酒。喝完酒有时又夹肉依次塞进对方嘴里，最后才松开扯耳朵的手。

揽颈酒

壮族揽颈酒可以在敬酒时进行，也可以在酒酣时进行。饮时两人并排站好，你揽我的颈，我揽你的颈，然后各人抽出一只手拿酒杯，你给我饮，我给你饮，同时饮下。一般只饮一杯，以表示手足情谊。

第三节　饮酒禁忌

酒文化是少数民族礼仪文化的重要组成部分，各民族在斟酒、敬酒和劝酒等方面都有一些规矩和禁忌。

斟酒禁忌

酒满为敬是各族共有的习俗。彝族谚语说："酒满敬人，茶满欺人"。哈尼族古歌唱道：喝酒要拿大碗，倒酒要倒满。吃菜要拿大碗，挟肉要挟最肥的一片。喝酒要喝出好音，喝酒的声音要像溪水流淌一样好听；挟菜要挟出好样，挟菜的筷子要像蝴蝶采花一样好看。[①]

在少数民族聚居区，不论是彝山苗岭还是壮家傣寨，亲友相聚，主人设杯置碗以酒相待，斟酒时一定要双手执壶将酒轻缓地斟入杯中，直至杯满，斟酒不满或是来宾中斟酒不均匀，都会引起宾客的不满。有经验的斟酒人，能利用液体的表面张力，徐徐注酒入杯碗，使酒液微微凸出杯碗口而不溢出，以示对宾客的敬重。若无外客，居家小酌的环境要相对宽松些，斟酒者可在征求饮用者意见的条件下适度控制斟酒量。

"旧的不去，新的不来"的观念在斟酒习俗中表现十分明显。彝族谚语说"老酒喝不掉，新酒烤不好"，哈尼人则说得更为形象"主人盛情的米酒，洗白了天空的脸。男人抬出的米酒，喝去了筒里的一点；女主人摆出的饭菜，吃去了碗里的一点。客人喝去的米酒，会随着蘑菇房上升起的炊烟，回到主人的木楼；客人吃去的饭菜，会随着山菁里腾起的白云，回到哈尼山寨"。

敬酒禁忌

彝族、苗族在宾客临门时，常以牛角杯、羊角杯或猪蹄杯敬酒。敬酒时不能把杯的角尖弯向客人，会被视为不友好。客人回敬时，也不能把角尖弯向主人，而应双手平举角杯，使角尖右弯或弯向自己。傣族敬酒饮酒，有把酒倾注少许在地上的习俗。据说，魔鬼们在酒碗边上抹上毒药，被聪明的人发现后，喝酒前先把满碗的酒拨出一点，以洗掉碗口上的毒汁，既喝了美酒又免除了祸害。所以到景颇山寨、傣家、佤山做客，如果遇到主人敬酒时或饮用前倾注少许在地上，无须惊讶，这正是主人对客人表示尊重的行为，也是少数民族在长期发展中一种自我保护的近乎本能的反映。

佤族群众也有许多敬酒禁忌。在阿佤山区，递酒给阿佤人时，双手要前伸，手心向上，绝不能手心向下、大拇指与四指分开。饮酒时，主人将酒淋几滴在地，先喝一口，才递给客人。客人双手接碗，以碗不离口一饮而尽，表示对主人的尊重；

---

① 西双版纳傣族自治州民族事务委员会编：哈尼古歌，云南民族出版社1992年版。

如果确实不会喝，也要抿上一口，再三向主人致歉，并由其他人向主人解释才行。阿佤人饮酒的另一习俗是，主客围蹲在地上，主人用右手把酒递给客人，客人也必须右手接酒，先淋少许在地或用手指弹一点在地上，意为祭祖，以表示对主人祖先的尊重，然后才能饮用。有饮用转转酒习俗的民族，共用一只碗或筒饮酒时，上家在饮酒后，要用手掌心在饮过的碗口或筒边轻轻抹擦一下，把沾在碗口或筒边的残酒抹干净，才双手递给下家，以示互相尊重。

劝酒禁忌

景颇族热情好客，凡是来的客人，主人都会热情招待。主人递来的烟酒，客人必须用双手去接。熟人间相互敬酒，不是接过酒来就喝，而是先倒回对方的酒筒里一点再喝，主人认为这是互相尊重之意。几个人一同到景颇人家，主人一般不亲自一一敬酒，而是把酒筒交给看上去年纪大点的人。如果把酒筒交给了你，说明把心都交给你了，意思要你代表他的心意，给大家敬酒。喝酒应用筒盖，不能拿酒筒直接喝。按照景颇人的习俗，来客不管是男是女，一进门就开始敬烟、敬酒。景颇族敬酒有许多特点，一种是客人到来，用竹酒筒直接倒给客人喝。一种是把酒杯放在火塘边，慢慢地敬献给客人享用。一种是一面与客人交谈，一面给人敬酒。无论是敬酒还是敬烟，景颇族一般是先敬长辈，后敬平辈，再敬后辈。

布依族还有饮酒劝歌的习俗。其形式有二人对唱、分组对唱、盘歌对唱等。在宴席迎亲时主人劝酒一杯，热情地先唱酒礼歌："贵客到我家，如凤落荒坡，如龙戏浅水，实在简慢多。"客人听后欣然举杯，唱"赞美歌"回敬："喝酒唱酒歌你唱我和，祝愿老人寿比南山；祝福后生下地勤做活；祝福姑娘家织布勤丢梭；祝福主人家年年丰收乐。"这样陪客的人也得一饮而尽。每饮一杯酒必对一曲歌，客人或陪客者对不上来，就要被罚喝"哑杯"。也有的入席后，一般总是由会唱歌的姑娘向客人敬酒，姑娘每唱一首，客人亦应回唱一首，唱不出者则罚酒一杯。在一片喝彩声中，被喝得昏昏欲醉。客人告辞，主人唱"送客歌"送至家门，大家情深谊重，依依不舍。

阿昌人很注重酒德，认为醉酒是丢脸的事。每当热情好客的主人多次斟酒劝酒时，许多明智的客人一方面向主人表示谢意，另一方面则用"酒在壶中人论酒，酒到肚中酒弄人"的民间俗语来回答主人，说明自己实在不能再饮酒了。主人听到客人的这些话后，一般也会停止劝酒。

第六章　深情款款祝酒歌

云南各少数民族不仅善酿酒，而且好唱歌。他们用歌来阐释本民族古老的历史以及本民族的生活方式、风俗习惯和对美好生活的向往。酒与各民族的日常生活密不可分，因而"美酒歌来伴、有酒必有歌"的习俗便赋予了酒歌更多的文化内涵，使其展现出各民族的性情、文化和传统。

酒歌又称为"酒礼"或"酒礼歌"。酒歌历史悠久、源远流长，歌词即兴而得，广泛运用对偶、比兴、夸张、排比等修辞手法，是各族人民智慧的结晶。酒歌的创作与传唱使酒与民间文学紧密结合，生动地反映各民族特有的生活方式、风俗习惯以及他们勤劳俭朴的高尚品德和美好的心灵。目前流行的少数民族酒礼歌分为：酿酒歌、敬酒歌、婚嫁歌、节庆歌、祭祀歌等。主唱客答，主客间的感情交流都包含在这些朴实大方，讲礼好客的酒歌中。

第一节　敬酒歌

云南各少数民族都热情好客，他们以香醇的美酒和甜美的歌声款待客人。酒香令人陶醉，歌美让人留恋。凡是到了少数民族村寨的客人都会收到这样热忱的邀请：

普米族迎客调

欢迎你啊朋友

我亲爱的朋友伙伴们，

来普米家花床上坐一坐。

花木碗里敬热茶，

玉石碗里敬牛奶，

金边碗里敬你一碗新酿的酥里玛，

银边碗里敬你一碗甜蜜的蜂糖。

普米人民最好客，

来普米家花床上坐一坐。①

傈僳族迎宾调

请到我村走一走，

请进我家坐一坐。

最甜的杵酒杵给你，

最香的蒸酒端给你。

一同跳跳仟俄，（仟俄：傈僳族舞蹈）

一起唱唱优叶。（优叶：傈僳语音译，傈僳族民间情歌曲调名）

请把快乐跳出来，

请把幸福唱出来。②

推杯换盏间，必要对酒当歌。除了迎客调，各民族还会唱敬酒歌、劝酒歌、谢酒歌、和留客调等。布依族爱酒，且能歌善舞，讲究"无酒不成礼，有酒必有歌"的待客形式；认为有歌有酒，才能表示待客的真诚热情，因此请客吃饭时都要唱酒礼歌。主人在歌中对客人的来临表示欢迎，为招待不周表示歉意；客人也以歌作答，对主人的热情款待表示衷心感谢，对美酒佳肴、烹调技艺进行赞美歌颂。

布依族敬酒歌

主人：说是酒席的桌面，

　　　桌上只有一碗荔头，

　　　席上是一碗萝卜，

　　　没有一片肥肉掺杂，

　　　没有一片肥肉混合。

客人：我在我家吃菜根，

---

① 普米族民间文学集成编委会编：普米族歌谣集成，云南民间文艺出版社1990年版。

② 云南民间文学集成办公室、保山地区民间文学集成小组编：傈僳族风俗歌集成，云南民族出版社1988年版。

> 走到您乡吃海参。
>
> 酒吃人情肉吃味，
>
> 多谢亲家好欢心。
>
> 我在我家吃辣椒，
>
> 走到您乡吃燕窝。
>
> 酒吃人情肉吃味，
>
> 多谢亲家好欢乐。[①]

即使满桌是山珍海味，主人也这么唱，以表达自谦；心存感激的客人则要表达对主人的褒扬和赞美。"酒吃人情肉吃味"则体现了酒的媒介作用，即，酒连接和凝聚的是一份浓浓的亲情友情。哪怕是简单的菜肴，但主人的盛情，胜过了山珍海味。

布依族酒席上，饮酒的每一个环节都是一个仪式，而仪式的重要表现就是唱歌。客人进家，要唱开财门歌，祝福主人家财源滚滚；酒席开始前，要唱解桌歌；坐上酒桌，要唱分筷歌；分完筷子，要唱解壶歌，等等。

主人：楠竹筷子一双双，

　　　妹来分给唱歌孃，

　　　妹来分给唱歌手，

　　　快快乐乐上歌场。

　　　楠竹筷子黄晶晶，

　　　妹来分给唱歌人。

　　　妹来分给歌师傅，

　　　快快乐乐上歌亭。

客人：楠竹筷子对对齐，

　　　姐来分筷我欢喜。

　　　姐姐分筷到我手，

　　　大家同陪这桌席。

　　　楠竹筷子黄金金，

---

① 杨有义编：布依族酒歌，贵州人民出版社 1988 年版。

姐来分筷笑盈盈。

姐姐分筷到我手，

妹来唱歌贺主人。①

　　席上吃酒莫推让（布依族劝酒歌）

主人：一张桌子四角方，

　　　七碟韭菜八碟姜，

　　　碗碗都是清汤水，

　　　没有哪碗是肉汤。

　　　请来的兵难打仗，

　　　田中无水难栽秧。

　　　席上吃酒莫推让，

　　　妹我无钱慢想方。

客人：我到姐家姐本贤，

　　　清炖红烧几大盘，

　　　顿顿吃的八宝饭，

　　　烟水茶果用不完。

　　　处处把我当客待，

　　　点点滴滴记心间，

　　　姐的深情厚谊重，

　　　就是吃水心也甜。②

　　壮族以交杯酒敬客人。敬酒时不用酒杯，而是用白瓷汤匙。两人从酒碗中各舀一匙，相互交饮。主人这时还会唱起敬酒歌：

　　　锡壶装酒白连连，

　　　酒到面前你莫嫌，

　　　我有真心敬贵客，

①② 杨有义编：布依族酒歌，贵州人民出版社1988年版。

敬你好比敬神仙。

锡壶装酒白瓷杯，

酒到面前你莫推，

酒虽不好人情酿，

你是神仙饮半杯。^①

布依族敬酒歌

主人：一杯酒，我来斟，

　　　难得姊妹做一堆，

　　　难得姊妹来相会，

　　　淡酒也要敬一杯。

客人：二杯酒，满满斟，

　　　金杯照见敬酒人，

　　　姐你诚心来敬酒，

　　　情也深来意也深。

主人：三杯酒，斟满缸，

　　　斟杯素酒劝客尝，

　　　酒不好来情义在，

　　　难得姊妹到我方。

客人：四杯酒，滴滴甜，

　　　主人情义重如山，

　　　自古良言说得好，

　　　仁义重来水也甜。

主人：五杯酒，斟满瓶，

---

① 中国歌谣集成云南卷编辑委员会编：中国歌谣集成（云南卷），中国 ISBN 中心、新华书店北京发行所 2003 年版。

我斟苦酒敬亲人。
敬杯苦酒表心意，
难得姊妹到我城。

客人：六杯酒，香喷喷，
香满歌亭香满城，
自古常言说得好，
人讲礼义值千金。

主人：七杯酒，酒味酸，
七姊七妹来团圆。
我拿酸酒敬姊妹，
不讲酒酸讲心甜。

客人：八杯酒，滴滴香，
主人斟酒劝客尝，
主家烤的桂花酒，
五湖四海美名扬。

主人：九杯酒，酒味辣，
斟杯辣酒敬姨妈，
我拿辣酒敬姊妹，
家无好酒有办法。

客人：十杯酒，斟得全，
十家姊妹大团圆。
主家情义深似海，
富贵荣华万万年。①

---

① 杨有义编：布依族酒歌，贵州人民出版社 1988 年版。

苗族敬酒歌

主人：大叔高龄见识广，

　　　敬你一杯表心肠。

　　　酒敬青年得对象，

　　　酒敬老人寿命长。

客人：虚度光阴几十年，

　　　从不出村见识浅。

　　　多谢姑娘敬我酒，

　　　祝你满意结良缘。①

　　彝族不仅善酿好饮，而且待客热情，总要奉上美酒，唱起富有民族特色的酒歌。敬酒、劝酒时要唱，送别客人时也要唱。

彝族敬酒歌

第一杯荞酒哟，敬先祖，

他们开天辟地挡洪水。

第二杯荞酒哟，敬长辈，

他们烧地种荞养儿女。

第三杯荞酒哟，敬兄弟，

祖宗家教要牢记：

"孔雀比锦鸡好看，

和睦比凤凰更珍贵。"

第一块坨坨肉给灾民，

他们丢了家，流下泪。

第二块坨坨肉敬穷人，

他们昼夜辛劳，流尽汗水。

第三块坨坨肉敬亲家，

---

① 中国歌谣集成云南卷编辑委员会编：中国歌谣集成（云南卷），中国 ISBN 中心、新华书店北京发行所 2003 年版。

"两家人共一堵墙，
墙是祖宗留下的。"①

彝族劝酒歌

想喝不想喝都要喝

阿老表，端酒喝

阿表妹，端酒喝

阿老表，喜欢不喜欢也要喝

阿表妹，喜欢不喜欢也要喝

喜欢呢，也要喝

不喜欢，也要喝

管你喜欢不喜欢

也要喝

管你喜欢不喜欢

也要喝

阿老表，端酒喝

阿表妹，端酒喝

哥有情，妹有意

要喝就喝到月亮落②

（重复）

米酒喝个够（彝族留客调）

亲戚你莫走哎，你莫走，

猪羊多多有哎，多多有，

杀了九小头哎，九小头，

还有九十九哎，九十九。

朋友你莫走哎，你莫走，

米酒多多有哎，多多有，

---

①② 云南民间文学集成办公室编：云南彝族歌谣集成，云南民族出版社1986年版。

喝了九小坛哎，九小坛，

还有九十九哎，九十九。

饭菜甜又香哎，甜又香。

米酒喝个够哎，喝个够。

亲朋坐拢来哎，坐拢来。

举杯庆丰年哎，庆丰年。[①]

第二节　婚嫁歌

云南许多少数民族中都盛行迎亲或婚庆唱酒歌的习俗，几乎婚庆的每个环节都要唱酒礼歌，如提亲歌、嫁女歌、哭嫁歌、送亲歌、接亲歌、祝愿歌等。如在布依族婚礼上，年长者饮酒后常在堂屋中唱起《酒歌》《开亲歌》和《古歌》等，以祝贺主人家；主人家也以酒、茶回敬客人们。

彝族姑娘出嫁时，有哭嫁的风俗。不仅姑娘自己哭，就连陪伴她的姊妹们都要哭。其间要唱哭嫁歌，听起来情真意切，催人泪下。婚礼当天，女方家在堂屋中摆上一张桌子，新娘和女方家本家族的姑娘们坐在桌旁，而本家族的媳妇们则站在桌子的两边唱酒礼——劝嫁歌。随着这凄楚的歌声，新娘由自己的弟弟背着在屋内围着桌子转三圈后，于一片哭声中将她背到门外。屋外火堆旁，妇女们接着唱起了酒礼——出嫁歌。此时，女方家拿出早已准备好的哑酒，将一根竹管插酒坛，人们将酒吸出注于碗中边唱边喝。

接亲队伍到女方家后，送亲先生要与接亲先生一起在大门旁合唱酒礼歌。酒礼歌的内容有人类的产生、六祖起源等，边唱边将新娘和送亲婆带到堂屋中，从左到右绕三圈，开始敬酒。新娘敬酒敬到接亲先生处，就在他面前跪下，向他要耳镯。这时，接亲先生边喝酒边用唱酒礼的形式讲述耳镯的来源、耳镯的作用等。之后，接亲队伍就把新娘接带回新郎家。

接亲队伍回到新郎家门口时，新郎家早已在院子里摆上桌子和酒菜，接亲先生和送亲先生要对唱酒礼，由接亲先生先唱，一问一答，一唱一和，表示迎亲。接着是认亲敬酒仪式。敬酒敬到叔子面前时，新娘要口衔毛巾，手端酒杯，双膝跪下。

---

① 云南民间文学集成办公室编：云南彝族歌谣集成，云南民族出版社 1986 年版。

送亲先生则一边扶着新娘，一边唱酒礼，意思是从今以后把姑娘托付给新郎家了，要叔婶们多方帮助，耐心指教。接亲先生立即接唱，表示接受和还礼。最后，两位先生又合唱酒礼，引导新娘今后如何为人，并代表父母给新娘以教诲。

彝族新娘认亲敬酒以后，亲戚们燃放鞭炮，会唱歌的又聚在一起，边喝咂酒边唱歌，以酒润喉，以歌助兴，欢乐喜庆的气氛又达到一个高潮。这天晚上，主人家就带着一些会唱歌的人，带上酒，向暂住在邻居处的亲戚敬酒。敬酒时，先唱敬酒歌，然后再唱挽留歌。亲戚们也唱歌作答，并谢承好意。这时，男方家院子里篝火熊熊，婚礼在歌声中和酒香中沉浸一夜。

粗母作候（彝族酒礼歌节选）

女歌手：两位押礼先生呃，

　　　　请不要忙着喝酒！

　　　　你们来时山上没有一条羊肠小路，

　　　　难道长出翅膀从云里飞走？

　　　　路上玉宇琼楼闪闪发光，

　　　　千万颗银针是否刺得你们眼泪流？

　　　　道路方向难辨好像涂了一层油，

　　　　不知你们摔了多少筋斗？

押礼先生：唱歌的表妹呃，

　　　　我来时冰天雪地，

　　　　哪个敢往羊肠小路上行走，

　　　　就踏着大路翻上山梁。

　　　　只见冰凌把山岭打扮成银色宫殿，

　　　　阴天走路雪地不会反光。

　　　　我放牛羊踩惯这条路，

　　　　闭着眼睛也认得出方向。

女歌手：两位押礼先生呃，

　　　　请不要忙着拣菜，

圈里有二百九十七只羊，

问你们分成几帮上山去放，

不准许哪帮多一只，

不准许哪帮少一双。

如果你们两个猜不着，

罚喝三大碗酒才算过关。

押礼先生：唱歌的表妹呃，

板栗破壳能见米，

圈里有二百九十七只羊，

分成三帮赶到牧场里，

一只不多一只也不少，

每帮羊子都是九十九。

拿起鞭子打自己的腿呀，

请表妹把三碗酒喝下去！ ①

《粗母作候》是彝族婚礼中女方家歌手给押礼先生出难题的对答歌。彝语"粗母"为押礼先生。彝语"作候"是吃饭礼节。押礼先生谙熟本民族风土人情，能歌善舞，代表男方家押送聘礼去女方家，要闯过多道关口，给男方家娶亲增添荣光。如果押礼先生在唱《粗母作候》中被女方歌手斗败，就会受到不给饭吃的惩罚，只好饿肚子。

苗族婚礼上，主客双方常邀请歌手对唱酒歌，客方歌手在堂屋，主方歌手在伙房，其余人边听歌边轮流饮酒。喝酒唱歌，唱歌喝酒，主唱客和，通宵达旦。

（苗族婚嫁歌）

哪个都想找会做针线的姑娘，

哪个都想找善酿美酒的姑娘。

会做针线的姑娘从"羊"的地方来，

善酿美酒的姑娘从云里来。

---

① 云南民间文学集成办公室编：云南彝族歌谣集成，云南民族出版社 1986 年版。

包了酒药从太阳照的地方来，

带了酒药从月亮亮的地方来。

……

伸出左手接住酒碗吧，

用右手送到嘴边喝起来。

大人们都说酒又香又好喝，

小娃娃们说酒好香好甜。

美名传到了远方啊，

美名传到了大地方！ ①

普米族从恋爱到说亲、娶亲、出嫁等整个过程都有歌声相伴。《情歌》是表现男女青年双方相互的爱慕，《认亲调》则是男方到女方家去相亲说媒。到了结婚喜庆，有男方到女方家迎亲时唱《迎亲调》《出门调》。接新娘时双方对唱《盘婚调》，新娘上马离家唱《上马调》，半路上遇着新郎的迎亲队时唱《下马调》，新娘接到男方家时主婚人唱《关门调》，欢迎四方亲友来祝贺唱《迎客调》《做客歌》等。这类歌曲表达的情感热烈而朴实，如《接亲调》唱道：

你已来到珊瑚玛瑙成堆的地方，

家中的老人对你有什么嘱咐？

珊瑚玛瑙做的装饰品最美，

要我背一箩回家装饰新娘。

你已来到珍珠成串的地方，

邻居们对你有什么嘱咐？

串串珍珠闪烁着耀眼的光芒，

要我选一串回家挂在新娘胸前。

你已来到银子闪亮的地方，

亲戚们对你有什么嘱咐？

闪亮的银子是值钱的东西，

要我背一袋回去买田买地。

① 中国歌谣集成云南卷编辑委员会编：中国歌谣集成（云南卷），中国 ISBN 中心、新华书店北京发行所 2003 年版。

你已来到金子闪光的地方，

家族们对你有什么嘱咐？

闪光的金子无比珍贵，

要我背一袋回去起房盖屋。①

（佤族婚嫁歌）

啊——

女大当嫁，

男大当婚。

你俩磨快了长刀，

你俩谈好爱情。

你俩在棉花树下谈情，

你俩在屋檐下说爱。

你俩一同上山打猎、摘果，

你俩一同下河摸鱼、戏水。

彼此诉明了衷肠，

倾诉了大山一样的忠贞。

山盟海誓结成对，

花好月圆共夫妻。

我们准备好了铓槌，

我们准备好了彩礼。

现在要办得合理合俗，

要把婚事办得清清楚楚。

你们可以育女，

你们可以生儿。

育得姑娘教她会买卖，

生得儿子教他会开田。

从此，让你们健康长寿，

---

① 普米族民间文学集成编委会编：普米族歌谣集成，云南民间文艺出版社1990年版。

从此，让你们安然无恙。

让你们活到眉毛开花，

让你们活到胡须发白。

啊——塞信①②

第三节　节庆歌

少数民族的节庆典礼中必有酒宴，有酒便要唱酒礼歌。如每逢节庆，普米山乡便充满着节日的热闹与欢腾。普米人围坐在火塘边，边唱歌边饮酒，到处洋溢着愉快的歌声。具有典型代表的年节歌主要有《过年歌》《祝节歌》《敬神调》等。年节礼仪歌的曲调一般是婉转而深情，富有歌唱性，如《春节歌》：

啊哩呀哩——③

新春佳节到人间，

青绿的松枝发芽了，

鲜艳的山花开放了，

春色染浓了山寨。

香烟在木房上飘扬，

火塘里闪着金光，

火塘边围挤着幸福的人儿，

在这美好的节日里，

我们把遥远的祖先怀念。

没有过去，

就没有现在。

没有开头，

就没有结尾。

没有真理和信念，

---

① 啊——塞信：佤语，有"接酒啰"之意，同时，又是呼告语，有呼唤、赞叹之意。

② 中国歌谣集成云南卷编辑委员会编：中国歌谣集成（云南卷），中国 ISBN 中心、新华书店北京发行所 2003 年版。

③ 啊哩呀哩：普米族民歌衬词，有"你唱我唱"之意。

就无法翻过重重高山。
没有勤劳和勇敢，
就不能越过穷苦的深渊。
靠真理和信念，
能达到黄金的彼岸。
有勤劳和勇敢，
西宁骏马能配上金鞍。

"酥里玛"酒香又醇，
你一碗来我一碗，
美酒飘香千山外，
畅饮开怀心欢畅。
"猪膘肉"味道浓，
你一坨来我一坨，
汗水换来美味香，
普米家里不愁吃。
喝一口热一遍，
吃一块细品味，
祖先的遗言似珍珠串。
祖先的话语，
融化在酒碗里，
照亮了如锦似玉的前程。

白天过去是夜晚，
辛勤换来欢乐。
节日的夜晚多快活，
篝火旁围起了欢乐的舞圈。
心不散月儿圆，
祖先的话儿当歌传。
笛声脆亮又婉转，

相亲的伙伴舞翩跹。

唱歌又跳舞，

普米欢乐庆新年。

普米祝福又歌赞，

真理和信念随火花在飞溅，

酒满碗，茶满罐，

欢跳"锅庄"迎春天……①

傈僳族的《串亲调》以"盼亲人"（主人）与"串亲人"（客人）对答咏唱的形式，生动而形象地描绘了节庆期间走亲访友途中遇到并克服种种艰难险阻的情境和思念亲友、盼望亲友相见的迫切心情以及最终亲友得以相聚的美好时辰。酒歌的结尾，"串亲人"和"盼亲人"分别这样唱道：

串亲人：今天是美好的日子，

今夜是吉祥的时辰，

我们得相见了，

我们得会面了，

会到真心盼亲的人，

见到诚心待友的人。

你铺下羊毛毡，

你摆好了竹编凳，

你酿的酒甜蜜蜜，

你煮的饭香喷喷，

舍不得吃的拿来吃了，

舍不得喝的拿来喝了。

我已经吃得很饱，

我已经喝得很醉，

感谢你啊，煮饭人，

多谢你啊，泡酒人。

---

① 普米族民间文学集成编委会编：普米族歌谣集成，云南民间文艺出版社1990年版。

讲不完思念的话，

叙不尽想念的情，

唱过多少思念的歌啊，

哼过多少想念的曲子，

我已经唱了好些歌，

我已经哼了不少曲，

该慢慢地转去了，

该悠悠地回家了。

流着眼泪与你别离，

淌着眼泪和你分开，

舍不得分也得分开，

舍不得离也得离别，

明年不老我再来访友，

明年不死我再来串亲。

盼亲人：你我相会不容易，

你我相见难上难，

何必走得这样急？

何必去得这么忙？

我们在人间度一世，

我们在世上过一生，

青春不会复返，

死后不能重生。

怎舍得这么快就分别，

怎舍得这样早就别离。

我的心真难过，

我的肝中多忧愁，

没法让眼泪不流，

无法让泪水不淌。

我年年把你等，

我月月把你盼，

我好像是离群的雀鸟，

我好比离枝的落叶。

再伤心也得分别啊，

再难过也得别离，

杜鹃在你身边唱，

带去我对你的思念，

阳雀在你耳旁叫，

带去我对你的想念，

望你明年再来访友，

盼你明年再来串亲。①

贺年词（彝族）

今天晚上呀，

我们要过年。

月花哪里来？

要算年，

要算月，

怎么算年月？

岩头有树木，

岩脚有藤子，

年头和年尾，

对木有数；

藤子有数，

年花是粮食丰收花，

月花是牲畜兴旺花。②③

---

① 云南民间文学集成办公室、保山地区民间文学集成小组编：傈僳族风俗歌集成，云南民族出版社1988年版。

② 年花：年份。月花：月份。

③ 云南民间文学集成办公室编：云南彝族歌谣集成，云南民族出版社1986年版。

每年秋收新谷成熟季节，许多民族都有尝新与喝新酿米酒的礼俗，俗谓"新谷酒"。布依族在秋季庄稼收获后，便开始忙碌着酿制新米酒，腌制腊肉了。待次年农历正月间农事不忙时，便邀请亲戚朋友到家中来欢庆丰收喝酒唱歌，称为丰收酒礼。这一天主人家便早早起床，准备好各种佳肴，等待客人的到来。客人陆续到来后，便被请到火塘边，大家围坐在一起，

融融的火焰，把房间照得暖洋洋。这时，主人便来敬酒，同时欢乐地唱起《客气歌》：

主人：昨晚灯花爆，

今早喜鹊叫，

都说要有客，

贵客真来到。

接着，主人便站起来打开酒坛，酒香四溢，客人闻后便唱道：

客人：米酒绿中央，

开缸十里香，

下河洗坛子，

醉倒老龙王。

主人谦虚地唱道：

美丽的孔雀，

刺笆林歇脚，

饿了没食吃，

渴了没水喝，

怠慢贵客了，

只是来空坐。

客人接着称赞道：

酒肉摆满桌，

八碗九缸钵，

压弯桌子面，

压断桌子脚。

就这样主客互相祝贺歌唱，直到散席。丰收酒的礼仪结束，主人送客人到寨边时，双方共同唱道：

蜡梅花谢了，

桃花正含苞，

吃过丰收酒，

分手把家回。

春风轻轻吹，

大地微微暖，

家家选好种，

户户积田肥。

阳雀声声催，

节令莫迟违，

春天早下种，

金秋谷成堆。

年年米酒绿，

岁岁猪羊肥，

世世享安康，

代代都富贵。[①]

唱罢，主客互相拱手道别，并相邀来年再相聚共饮"丰收酒"。

第四节　祭祀歌

云南各少数民族不仅将酒用于各种祭祀活动，而且将酒歌与祭祀礼仪相结合，吟唱祭祀酒歌。根据祭祀内容和祭祀礼仪，祭祀酒歌分为祭祖歌、驱邪祭祀歌、狩猎祭祀歌、农耕祭祀歌、年节祭祀歌、生活祭祀歌等。

居住于哀牢山中的拉祜族苦聪人，世代以狩猎为生。他们称猎神为"沙尼"，每次出猎，猎人们都要杀鸡祭祀猎神，并唱"猎神调"：

今天，我杀鸡来祭你，烧着香火来献你，

煮饭来祭你，磕头来祭你。

从今后，你给我进得了山，你给我钻得了菁，

---

① 杨有义编：布依族酒歌，贵州人民出版社 1988 年版。

你给我过得了河，你给我爬得了坎，

你给我歇得了夜，你给我打得到野兽，

你给我一年四季都有吃。

进山，给我看得见野兽，不要给我放空枪，

不要给我射空箭，不要让野兽伤害我。

打得野猪献给你，打得麂子献给你，

打得白鹇献给你，打得野鸡献给你，

无论打得什么，都是你的，我都献给你。①

祭猎神词（彝族）

天阿爸，请饮酒！

地阿母，请饮酒！

皮武妥，请饮酒！（彝族传说中开天之神）

列哲社，请饮酒！（彝族传说中辟地之神）

水府三官，请饮酒！（水中各种神的总称）

先知先觉的神灵啊，请饮酒！

博闻广见的神灵啊，请饮酒！

山上的山神，岩上的岩神，请饮酒！

白岩上的白岩神，红岩上的红岩神，请饮酒！

山顶、山腰和山脚，三级树神请饮酒！

山顶、山腰和山脚，三级洞神请饮酒！

岩峰犀牛神，

山顶青蛇神，

山腰灰蛇神，

山脚花蛇神，

水面的鱼神，

水底的龙神，

路边四脚蛇，

---

① 孙敏等编：拉祜族苦聪人民间文学集成，云南人民出版社 1990 年版。

桥头花鱼精，

箐中白角獐，

林中白角麂，

诸位神灵，一同敬请！

敬请天显灵，天神管太阳，日出云雾散，大地光灿灿；

敬请天显灵，天神管星月，月亮明晃晃，星宿亮晶晶。

敬请地显灵，地神管草木，大树莫遮鸟，深草莫藏兽。

祭献众神灵，保佑打猎人，步步都平安，处处都吉祥！

祭献众神灵，请搜集走兽，请收集飞禽；

洞中的出洞，林中的出林；

弩箭出弓，百发百中。

打猎祭词，句句显灵！[①]

佤族春播前要祭"木依吉"，佤族原始宗教中掌管着风雨、生死和丰收的最高神。祭祀时要杀猪剽牛，并唱祭祀歌祈求社神的保佑：

寨上的社神啊，

寨下的河灵。

白露花已经开白了山坡，

刺桐花已经开红了寨周。

山风呼啸，

骄阳似火。

是我们播种的时候了，

是我们栽插的季节了。

我们要撒小米了，

我们要播稻谷了。

是白鹇带给我们的稻种，

是山雀带给我们的小米籽。

让它们落到地上，

---

① 云南民间文学集成办公室编：云南彝族歌谣集成，云南民族出版社 1986 年版。

让它们进到土里。

陡坡也能扎根，

岩上也能发芽。

小雀来了遮住它的脸，

松鼠来了捂住它的嘴。

让它蓬大，

让它叶壮。

让它穗穗饱满，

让它丘丘倒伏。

我们已经备仓以待，

愿我们收获昭昭。[①]

农耕祭祀歌（拉祜族）

河边鹅毛尾花开了，

山上麻毛尾花开了。

哥采一把祭山神，

妹摘一把献地神，

求神灵保佑谷花像索尾麻一样旺盛，

谷穗像娃阿得果一样丰满。

求神灵保佑开荒种地得丰收。

求神保佑谷米快快长大，

雀不来吃，虫不来吃，水不来冲，

人吃不完，畜吃不完，房堆不完。[②]

祭祖歌（彝族，节选）

人死三个魂，

一个随祖去。

---

① 秦家华、李子贤、杨知勇编：云南少数民族生活习俗志，云南民族出版社 1990 年版。

② 云南拉祜族民间文学集成编委会编：拉祜族民间文学集成，中国民间文艺出版社 1988 年版。

随祖这个魂，

供在香案上。

用草做祖身，

马樱花做手脚，

山竹做骨骼，

涂上黄颜色，

敬供灵台上，

儿子来献酒，

女儿来献饭。

……

你的子孙们，

绕纸又献饭，

祖灵来吃饭吧！

祖灵来喝酒吧！

年年祭祖灵，

祖灵保儿孙清吉，

保六畜兴旺，

五谷丰登。[1]

苗族的《芦笙词》是为亡人开路的，但几乎每一句话，都离不开一个酒字，说明酒在为亡人送行中很重要。在苗族人的观念里，如果死者得不到酒喝，其灵魂便得不到安抚，便不会平安地离去。所以，为亡人送行，是用敬酒的方式来完成的。

啊！亡人，现在你在等谁呀？

在等你的子女们来敬酒吧？

他们已给你敬七次酒了。

亡人！你吃不完就拿罐子来背起走，

你要拿这样的酒去敬祖宗。

亡人！家族亲友敬的酒你要背起走，

① 云南民间文学集成办公室编：云南彝族歌谣集成，云南民族出版社1986年版。

亡人快上路！

亡人要等亲人来敬酒。

亡人！快起来走吧！

你一定要等三亲六戚来拉你的手你才走吗？

现在家族的人已向你敬九回酒了。

你吃不完要拿罐罐来背起走，

要拿去送给我们的祖宗。

招魂词（拉祜族苦聪人）

哎，扎勒！

哎，扎勒！

舂米，米让它够煮饭，饭让它够；

煮菜，菜让它够。

无论做什么，都让它多多有余；

无论做什么，都让它顺心如意。

过来的日子，

做酒，酒不好；

煮饭，饭不够；

舂粑粑，也不会粘。

今天饭魂水魂，回家来，

做酒，让它酒好；

煮饭，让它吃不完；

舂粑粑，让它会粘。

煮菜，让菜够；

筛面，让面够；

筛米，让米够。

做魂回家走，

水魂饭魂一起回家来。①

---

① 中国歌谣集成云南卷编辑委员会编：中国歌谣集成（云南卷），中国 ISBN 中心、新华书店北京发行所 2003 年版。

参考文献

［1］ 普忠良编：中国彝族（《中华民族全书》杨宏峰主编），宁夏人民出版社
2013年版。

［2］ 关凯编：中国满族（《中华民族全书》杨宏峰主编），宁夏人民出版社2012
年版。

［3］ 唐洁编：中国德昂族（《中华民族全书》杨宏峰主编），宁夏人民出版社
2012年版。

［4］ 梁庭望编：中国壮族（《中华民族全书》杨宏峰主编），宁夏人民出版社
2012年版。

［5］ 韦学纯编：中国水族（《中华民族全书》杨宏峰主编），宁夏人民出版社
2012年版。

［6］ 杨春编：中国拉祜族（《中华民族全书》杨宏峰主编），宁夏人民出版社
2012年版。

［7］ 和向东编：中国普米族（《中华民族全书》杨宏峰主编），宁夏人民出版社
2012年版。

［8］ 陈国庆编：中国佤族（《中华民族全书》杨宏峰主编），宁夏人民出版社
2012年版。

［9］ 杨将领编：中国独龙族（《中华民族全书》杨宏峰主编），宁夏人民出版社
2012年版。

［10］ 李绍恩编：中国怒族（《中华民族全书》杨宏峰主编），宁夏人民出版社
2012年版。

［11］ 木仕华编：中国纳西族（《中华民族全书》杨宏峰主编），宁夏人民出版社
2012年版。

［12］ 潘琼阁编：中国瑶族（《中华民族全书》杨宏峰主编），宁夏人民出版社
2012年版。

［13］ 杨伟林、张云霞、王锋编：中国白族（《中华民族全书》杨宏峰主编），宁夏人民出版社2012年版。

［14］ 熊顺清编：中国阿昌族（《中华民族全书》杨宏峰主编），宁夏人民出版社2012年版。

［15］ 陶玉明编：中国布朗族（《中华民族全书》杨宏峰主编），宁夏人民出版社2012年版。

［16］ 周国炎编：中国布依族（《中华民族全书》杨宏峰主编），宁夏人民出版2011年版。

［17］ 苏发祥编：中国藏族（《中华民族全书》杨宏峰主编），宁夏人民出版社2011年版。

［18］ 祁德川编：中国景颇族（《中华民族全书》杨宏峰主编），宁夏人民出版社2011年版。

［19］ 欧光明编：中国傈僳族（《中华民族全书》杨宏峰主编），宁夏人民出版社2012年版。

［20］ 何少林、白云编：中国傣族（《中华民族全书》杨宏峰主编），宁夏人民出版社2011年版。

［21］ 李泽然、朱志民、刘镜净编：中国哈尼族（《中华民族全书》杨宏峰主编），宁夏人民出版社2011年版。

［22］ 陈国庆、谢玲：中国少数民族风情游丛书　佤族，中国水利水电出版社2004年版。

［23］ 白兴发：彝族文化史，云南民族出版社2014年版。

［24］ 杨学政编：云南少数民族礼仪手册，云南民族出版社1999年版。

［25］ 王忠华编著：独龙族，吉林出版社2010年版。

［26］ 云南省民间文学集成编辑办公室编：云南彝族歌谣集成，云南民族出版社1986年版。

［27］ 杨有义编：布依族酒歌，贵州人民出版社1988年版。

［28］ 李晓岑：云南科学技术简史，科学出版社2013年版。

［29］ 王忠华编：独龙族，中国文化知识读本，吉林文史出版社2010年版。

［30］ 天龙编：民间酒俗，中国民俗文化丛书，中国社会出版社2006年版。

［31］ 萧家成：中华民族酒文化——升华的魅力，华龄出版社2007年版。

［32］ 何明、吴明泽：中国少数民族酒文化，云南人民出版社1999年版。

［33］ 赵朕、赵叶、鲁保中、吴瑞云编：少数民族风情，中国旅游出版社2006年版。

［34］ 毛艳、洪颖、黄静华编：西南少数民族民俗概论，云南大学出版社2012年版。

［35］ 梁玉虹编：云南民族食俗，云南科技出版社2017年版。

［36］ 中国歌谣集成云南卷编辑委员会编：中国歌谣集成（云南卷），中国ISBN中心、新华书店北京发行所2003年版。

［37］ 普米族民间文学集成编委会编：普米族歌谣集成，中国民间文艺出版社1990年版。

［38］ 云南民间文学集成办公室、保山地区民间文学集成小组编：傈僳族风俗歌集成，云南民族出版社1988年版。

［39］ 张穗、杨世全主编，大理州州志编纂委员会办公室编：白依源流及习俗，大理方志通迅，1996年2月。

［40］ 孙敏等编：拉祜族苦聪人民间文学集成，云南人民出版社，1990年版。

［41］ 梁庭望：壮族风俗志，中央民族学院出版社1987年版。

［42］ 余嘉华主编：云南风物志，云南人民出版社1991年版，第439页。

［43］ 红河哈尼族彝族自治州民族志编写办公室编：红河哈尼族彝族自治州民族志·苗族，云南大学出版社1989年版。

［44］ 楚雄市民族事务委员会编：酿酒的故事，楚雄民间文学集成，楚雄市民委1988年。

［45］ 云南省民间文学集成办公室编：水酒的来历，佤族民间故事集成，云南人民出版杜1990年版。

［46］ 彭义良：怒族朝山节，民族调查研究1986年第1期。

［47］ 王震亚编：什撰何大祖，普米族民间故事，云南人民出版社1990年版。

［48］ 苏胜兴等编：密洛陀，瑶族民间故事选，上海文艺出版社1980年版。

［49］ 萧家成翻译整理：勒包斋娃——景颇族创世史诗，民族出版社1992年版。

［50］ 艾获、诗恩编：佤族民间故事，云南人民出版社1990年版。

［51］ 云南省少数民族古籍整理出版规划办公室编：哈尼阿培聪坡坡，云南民族出版社1986年版。

［52］ 西双版纳傣族自治州民族事务委员会编：哈尼族古歌，云南民族出版社1992
年版。

［53］ 云南省民间文学集成办公室编：云南摩梭人民间文学集成，中国民间文艺出
版社1990年版。

［54］ 秦家华、李子贤、杨知勇编：云南少数民族生活习俗志，云南民族出版社
1990年版。

［55］ 云南拉祜族民间文学集成编委会编：拉祜族民间文学集成，中国民间文艺出
版社1988年版。

［56］ 中国网http://www.china.org.cn/china/index.htm

后 记

在教授西方语言文化的过程中，我也在思考如何避免语言教学中的文化生态失衡和中国文化失语，如何提高青年一代向外传播中国文化的能力等问题。为了让更多青年了解云南少数民族文化，增强其对中华文化和各民族文化的热爱之情和自豪感，我在参与了"英汉双语世界民族文化系列读本"《云南少数民族民俗文化概览》的编译工作之后，继续参加"英汉双语云南少数民族经典文化概览丛书"的编译工作。《竹筒里流淌的酒歌》可以帮助读者从酒文化这个窗口看到文化的多元性，提升他们的多元文化意识和传扬民族文化的使命感，提高他们讲好中国故事和民族故事的能力。本人负责绪言、第一章、第二章和第六章的编译工作，杨亚佳老师负责第三章、第四章和第五章的编译工作。

本书承云南民族大学外国语学院支持，承李强教授为本书作序，师友和家人也热情鼓励，在这里一并表示深切的谢意。

王睿

2021年10月

Introduction

Cultural Characteristics of Wine

Wine is not only a product, but also a symbol of the development of human material civilization. From the perspective of natural science, wine is a kind of beverage that contains the organic compound ethanol (namely alcohol) and produces various chemical effects on the human body. From the perspective of social science, wine is a kind of material culture that impacts on people's spirits, affects their behaviors, and permeates other aspects of human cultures.

The origin of wine in China can be traced back to ancient times. Both Oracle Bone Inscriptions and Bronze Inscriptions have the word "wine". "The Book of Songs" contains verses related to wine, such as "Harvest rice in October, make spring wine good and fragrant, for longevity of the master" and "Intoxicated with your wine and satisfied with your sincerity". "Historical Records · Yin Benji" includes descriptions about King Zhou's luxurious life, "A pool of wine, a forest of hanging meat, for feast on the revelry all night long." All these verses and records show that the origin of wine in the land of China has a history of more than 5,000 years.

In the long course of our history, Chinese people have continuously created various unique brewing techniques. In addition, China's vast territory provides different natural conditions in terms of geology, water resources, climate and products, which helped to increase the varieties of wine, showing the regional characteristics of wine culture. Different wines often come from different nationalities, reflecting different national cultures from specific perspectives. It can be said that in the history of Chinese civilization, wine culture is an important part of the traditional culture of various nationalities.

Although wine belongs to the category of food culture, its cultural connotation and cultural significance are not limited to food culture itself, but have gone beyond the scope of material culture and penetrated into many fields of custom culture and spiritual culture. In terms of material culture, according to their living environment, mode of production and living habits, various ethnic groups have selected different wine-making raw materials, created different wine-making techniques, and produced different wines. Classified according to the types of raw materials, there are grain wine, fruit wine and mixed wine. Categorized according to the wine-making techniques, there are steamed wine, fermented wine, and mixed wine. Divided by alcohol content, there are high degree, Medium degree and low degree wine. In terms of customs, the cultural connotation of drinking customs is more colorful. Regarding when to drink, when not to drink, how to drink, how to toast, the meaning of a certain way of drinking, etc., each ethnic group has formed its own unique customary regulations, which contain complex and profound cultural connotations. In terms of spiritual culture, drinking or prohibition, wine in festival celebrations, wine in life etiquette, as well as wine-related literature, art, beliefs and rituals, etc., all these are proclaiming the values, ethics, aesthetics and religions of various ethnic groups. Therefore, the connotation and significance of wine culture involves all aspects of the national culture system.

Wine Culture

The invention, production, circulation and consumption of wine and the resulting wine culture are closely related to all aspects of human production and life, thus forming a multi-level cultural network related to wine. Structurally speaking, wine culture is a comprehensive culture. It is mainly manifested as a liquid material culture, but if you look at the overall structure of brewing, drinking, and social functions, it also includes solid material culture, technical culture, custom culture, spiritual culture, psychological culture, and behavioral culture. The comprehensiveness of wine culture has built foundations and opened up new ways for us to study other social and cultural phenomena. Researchers can use the window of wine culture to make new discoveries in the fields of national politics, social economy, production and life, history, archaeological excavations, etiquette and customs, marriage and family.

Wine culture is national. First of all, wine culture embodies the unique material culture, spiritual culture and custom culture of different nationalities. Secondly, the different brewing materials, techniques, utensils, drinking methods and wine usage customs also have obvious national characteristics. In addition, although wine is widely used in the real life of various nationalities, the specific connotation and form of wine usage customs have their own national characteristics.

Wine culture is pervasive. Although wine culture is only a small part of the national traditional culture, it is full of vitality and strong permeability. On one hand, wine plays a specific role on various occasions such as religion, health care, politics, military affairs, festivals, and daily life. On the other hand, wine is also combined with various cultural phenomena to form some wine-related cultures. For example, Jiu Ling is a combination of wine and literary forms such as poetry and lyrics. Wine song and wine dance are the combination of wine with music and dance. Drunken fist and drunken stick are the combination of wine and traditional sports.

Wine culture is the product of civilized society. We can find the trajectory of the development of civilized society among the various phenomena of wine culture. The wine culture of ethnic minorities can provide samples for the study of the history of social civilization. Take wine utensils for instance. Nowadays, some ethnic minorities are still using wine utensils made of natural materials such as animal horns, animal bones, poultry claws, animal feet, bamboo, and wood. Such utensils demonstrate the unique cultural traditions of ethnic people and are of great value for studying the history and culture of various ethnic minorities.

Wine Culture and Culture of Ethnic Minorities in Yunnan

The creation of any ethnic culture depends on a specific natural geographical environment. Only when standing on the stage of nature can people engage in cultural creation activities, and the ethnic minorities in Yunnan are no exception. Their unique mode of production, life style, distinctive ethnic culture and wine culture are characteristic of local natural living conditions such as terrain and climate.

In such a land of various topography, changeable climate and abundant natural resources, are living people of 26 nationalities, 25 of which are ethnic minorities, including Yi, Bai, Hani, Zhuang, Dai, Miao, Lisu, Hui, Lahu, Wa, Naxi, Yao, Tibetan, Jingpo, Bulang, Pumi, Nu, Achang, Deang, Jinuo, Shui, Manchu, Mongolian, Buyi and Dulong. Each of these ethnic minorities has a population of more than 6,000 and a certain habitation area. Due to differences in ethnic origin, living environment, historical evolution, and social development, the ethnic groups thriving in this land have developed their respective unique styles for things ranging from modes of production, living customs, ethnic festivals, weddings and funerals to religions, literature, art, science and technology. Using different ecological adaptation methods, they have solved a series of problems such as clothing, food, housing, and transportation, and have inherited a variety of colorful ethnic cultures.

Yunnan ethnic minorities have a long history of wine-making and drinking. During the Warring States period, the wine-making skills of the ancestors of Banxunman had reached a very high level. The "sake" brewed by Banxunman was the best of the time. This ancient and traditional brewing technique is still the main brewing method of the famous Japanese "sake". "Yunnan Pictures and Scriptures" records that in the Three Kingdoms period, Achang people had colorful wine culture. They not only grew sorghum to brew wine, but also sang and danced while drinking. At the beginning of the Yuan Dynasty, the famous Italian traveler Marco Polo traveled to Yunnan. In "The Travels of Marco Polo", he recorded the situation of wine industry in Yunnan. When Xu Xiake traveled to Yunnan, he recorded a village specialized in the manufacture of "liquor medicines" (distiller's yeast), which indicates that the wine-making industry has developed to a considerable scale in the Ming Dynasty.

Most ethnic groups have explored a set of unique wine-making methods and techniques in their long-term practice, and have produced a variety of wine with different flavors. For example, Xiaoguo wine of Yi people, Menguo wine and purple rice wine of Hani people, Shutou wine of Dai people, glutinous rice wine of Buyi people, horse milk wine of Mongolian people, highland barley wine of Tibetans, Jiuqian wine of Shui people, medicated wine and snow pear wine of Miao people, Gudu wine of Nu people, distilled wine of Lisu people, Dongzong (Caryota urens Linn) wine of Lahu people, Jade wine of Bulang people, Sulima wine of Pumi people and glutinous rice wine of Dulong people, etc. As drinking has gradually become an important part of the daily life of ethnic minorities, wine is indispensable on occasions such as farming festivals, ceremonial rites, weddings, funerals, sacrificial rites, visiting relatives and friends, entertaining guests, and cooking.

Wine is an important means for maintaining social relations among ethnic minorities. All ethnic minorities living in Yunnan, except for the Hui ethnic group, have a traditional custom of drinking. Whether in material life or in spiritual life, wine is always indispensable. This could be seen in old sayings like "Feast without wine is not true feast." For many ethnic minorities, the etiquette of reciprocity and communication is condensed in wine. Wine has become a gift when visiting friends, a sign of respect when entertaining guests, a token of affection when expressing gratitude, an expression of sincerity when resolving hatred, and a bond for better interpersonal relationships. Whenever guests come to visit, the Baiyi people from the Yi ethnic group will treat them with rice wine and white wine. The generous host would invite guests to drink to their heart's content. In the Yi ethnic community of Luquan, Yunnan, when relatives and friends are full and ready to leave the table, the host would sing "Song for Retaining Guests" to persuade guests to stay long and express the host's sincere friendship towards guests: "I have a lot of wine. Though nine small jars of wine are drunk up, there are still ninety-nine jars." Drinking "Tongxin wine" is the main method for many ethnic groups to express emotions, enhance friendships, and eliminate estrangement. (Tongxin means being of one heart and one mind. When drink Tongxin wine, two presons hold one bowl and drink from the same bowl together to indicate that the two are of one heart and one mind.) Therefore, friends drink Tongxin wine for deeper friendship; couples drink Tongxin wine for their life-long commitment; people with divergent ideas drink Tongxin wine for resolution of their dispute.

Wine is the main component of the festivals of various ethnic minorities. When festivals come, people gather together for grand celebrations. They raise wine glasses to create a jubilant atmosphere, and forget about the hard work in ordinary days. Mu Nao Zong Ge is a grand festival of the Jingpo ethnic group. On the day when the festival is observed, Jingpo people gather together from near and far. The bamboo barrels on their backs are full of home-brewed wine, which is not only an important sacrifice to ghosts and gods, but also the main drink of the participants. Nu people start to hold the Flower Festival (also known as the Fairy Festival) on March 15th of the lunar calendar every year. The length of the festival is determined by the amount of wine. It won't come to an end before the prepared wine is drunk up. Certain ethnic festivals are simply named after wine. For example, the festival

celebrated by Hani people in the middle of March of the lunar calendar is called "drinking rice wine." During the period of drinking rice wine, each village will hold a banquet called "Zi Ba Duo", which means taking turns to entertain guests with wine.

Wine is an important symbol of life journey. In the celebrating rituals for the development of life course, various ethnic minorities often use wine as a symbol of life moving from one stage to another. Wine accompanies them through the stages of birth, marriage, and death. For example, when their first-born is one month old, Buyi people will hold a banquet to celebrate the birth of the child, which is called "First Month Rice Wine". Yao people have the tradition of drinking wine at the New Year party, which is held on the first day of the first lunar month. Families which have new babies would bring to the New Year party a pot of rice wine, two pieces of tofu, and a piece of pork to declare the birth of their new family members. Getting married is one of the major events in life. Although the marriage etiquette of different ethnic groups is different, wine is a must in the whole process from proposal, engagement to wedding ceremony. For example, Yi people "eat mountain wine" when choosing a spouse, and drink "horn wine" when welcoming their relatives to weddings; Sani people drink "joy wine" or "eat small wine" when they propose and engage in marriage; Lisu people and Nu people drink "Tongxin wine" when they hold weddings; Jinuo people drink "divorce wine" when they get divorced. In funerals, wine is also indispensable. After the death of their relatives, many ethnic minorities will invite elderly people in the village to open the granary for them, so that they could brew wine to entertain those who come to the funeral. In the funerals of Achang people, wine is used throughout the entire process, including seeing off the souls of the deceased, worshiping the earth god before the burial, and worshiping the mountain god during the burial.

Wine is also an important medium for ethnic minorities to express and meet their emotional needs. Wine at celebrations, weddings, and gatherings expresses joyful feelings, while wine at funerals of relatives and friends or in difficult situations could dispel sorrows. The singing and dancing art of ethnic minorities is also closely related to wine. As an ancient song of Yi people sings: "The etiquette of Yi family is to drink wine and sing." The wine songs sung by various ethnic minorities in Yunnan have become jewels in the treasure house of folk literature. Rhetorical methods such as exaggeration, personification, and metaphor are often used in wine songs to express emotions. Wine songs are rich in content, often describing life related to wine and expressing the emotions inspired by wine. Their vivid language is full of breath of life. Their mellow and harmonious rhythm is pleasing to ears. Wine arouses the zeal to sing and dance, which in turn brings enthusiasm to enjoy more wine. When integrated with folk song and dance art, wine displays fully the life, customs and artistic talents of various ethnic groups.

In summary, the wine cultures of the various ethnic groups in Yunnan are reflecting their regional and ethnic characters while amplifying the glory of Chinese wine culture. Just as a drop of water can reflect colorful sunlight, the wine cultures of ethnic minorities have shown us the panorama of ethnic cultures from new perspectives. Following the aroma of wine, we will find treasures while exploring the connotation and characteristics of ethnic minority cultures.

Chapter One The Mellow Fragrance of Wine in Daily Life

Influenced by the specific living environment, social and historical background, and cultural heritage, wine has taken an extremely important position in the daily life of ethnic minorities. It has become a bridge and bond of interpersonal communication. It is inseparable from farming festivals, welcoming guests, socializing, and gatherings with relatives. Household issues, friendship between hosts and guests, harmony among neighbors, growth of youth, and wishes of longevity to elders are usually talked about over a bowl of wine. In another word, there is "no feast without wine" and "no hospitality without wine".

Section 1 Wine and Life

Wine as a companion in daily life

The ethnic minorities in Yunnan have a long history of drinking wine in daily life. Ever since the early Ming Dynasty, the Yi people living in the Ailao mountains have been fond of drinking wine and singing songs. During the first and second months of the lunar calendar, when the spring blossoms were blooming, the youth of Yi "brought wine into the mountains..., drank while appreciating the moon, and did not return until dawn" ("Yunnan Chorography" Volume 185). During the Ming and the Qing Dynasties, the Yi people in the Jingdong area of southern Yunnan had a saying that "you can skip the meal but not the wine". The Lisu people in the Grand Canyon of Western Yunnan and on the banks of the Nu River used to "live on the top of the sheer precipice and overhanging rocks. They reclaimed the mountains and planted grains which were mostly brewed into wine. They enjoyed wine to their heart content day and night. When grains and wine were consumed up, they left with stiff bows and sharp arrows, both men and women, to hunt among the steep cliffs, running as fast as cunning rabbits" (Yu Qingyuan, "Weixi Travelogue"). Lisu people love wine. When they become "excited after drinking wine, men and women danced in circle, hand in hand, stamping their feet to the melody of the LuSheng (a reed-pipe wind instrument)"(Qing Dynasty, "Lijiang District Chronicles"). The ancestors of the Dai nationality "enjoyed liquor while males played the pipa and females the flute for fun" (Qing Dynasty, "Pu'er District Chronicles" Volume 18 "Ethnography"). After the harvest, Naxi people "sewed clothes, made wine, and enjoyed meals and wine indulgently" (Yu Qingyuan, "Weixi Travelogue"). Achang people "planted grains to brew wine, and would sing and dance whenever they drank wine" (Xie Zhaohu, the Ming Dynasty, "Yunnan Travelogue").

The inherited drinking customs have permeated all aspects of their lives. Many ethnic minorities refer to "drinking wine" as "eating wine". For example, Nu people eat wine when they are doing farm work, relaxing at home, and traveling. Wine is a must when entertaining guests. If wine is not offered, guests would feel being neglected. Yi people love wine. For them, "wine is enough for a feast", "without wine, there is no courtesy". In Yi families, respect is shown with wine. Whenever guests come, there is no tea ceremony, but a custom of pouring wine for guests. At weddings, Yi people regard "be satisfied with wine" as being respected, while satisfaction with the food is secondary. At funerals, the one who brings the most wine is regarded as the most respectful to the deceased. The folk proverb of the Yi nationality also says, "wine for household reconciliation, wine for wedding ceremonies, wine for New Year celebrations, wine for tiger month Torchlight Festival, wine for worshiping ancestors in goat month, wine for marrying daughters, wine for negotiations, wine for crusades, wine for chatting with neighbors, wine for banquets, wine for farming and grazing, and wine for visiting relatives outside the mountains." It can be seen that wine is closely connected with all aspects of Yi people's life.

Drinking is an important activity in the daily life of ethnic minorities. The Baiyi people, a branch of the Yi ethnic group living in Heqing County, western Yunnan, regard wine as the venerable. At gatherings with friends and relatives, respect and love are expressed with wine. Toasting would not stop until all are drunken with wine. When Baiyi women go out or work in the mountains, they often bring a bottle of rice wine and drink in case of thirst.

Baiyi people are warm and hospitable. Whether old friends or new friends from their own ethnic group or other ethnic groups, as long as they come to Baiyi people's house, Baiyi people will treat them with rice wine and liquor generously.

Lahu people, men and women of all ages, love wine. Whenever they drink in group, they would dance to the LuSheng to their heart's content. Wine symbolizes auspiciousness and jubilation in the social life of the Lahu ethnic group. It is used in many aspects of their life, such as weddings and funerals, festivals and ceremonies, mediating disputes, courteous social communication, mutual assistance, treatment of diseases, and production. It plays a role in mediating communication, promoting relationships, strengthening courage, dispelling pain and eliminating fatigue. The Kucong people of the Lahu ethnic group also uphold wine. One of their criteria for judging a person's virtue and ability is: "If you don't drink more than three bowls of wine at a time, you can't be considered a hero, and if you don't eat three bulks of meat, you can't be considered a capable person." Kucong's wine song sings:

My wine jar,

Placed like a pile of rocks,

Densely packed.

My wine bowls,

As many as mushrooms,

Layers upon layers.

My wine,

Flowing like spring water;

My rice wine,

As fragrant as Qilixiang blossoms.

So many wine jars,

I can't hold them all by myself;

So many wine bowls,

My family can't hold them all.

So much rice wine,

My family will not drink alone;

So much joy,

My family will not enjoy alone.

It is precisely based on this group consciousness of sharing sufferings and happiness with everyone, that a sincere and fiery sense of trust among the Kucong people is built. Even if foreign guests come to the Kucong cottage, the Kucong people would treat them sincerely with all they have at home:

Drink,

Drink it happily!

My heart is as mellow as the rice wine;

Chew,

Chew full!

My heart is as warm as the fire pit.

Jingpo people, men or women, young or old, would always carry a small bamboo wine tube in their shoulder bags when going to market, visiting relatives and friends, offering bulls to ghosts, celebrating weddings or holding funerals. Whenever guests come to visit, or confidants meet, they will pass their bamboo wine tubes to each other. The person who receives the wine tube would pour a cup, firstly for the elderly, secondly for the person who passes it, to show courtesy and respect for each other. They generally regard wine as a kind of delicious food, so meals are not served without wine. In the life of Nu people, drinking is regarded as the happiest thing to do. After dinner, Nu people would gather together, drink wine, talk about wine, and express their feelings and thoughts. Usually, this is the greatest moment of the day.

Dai men drink at almost every meal. To Dai people, wine is indispensable for festivals, entertainment, religious services, weddings, funerals, etc. They say that if there is no wine in life, it is like a feast of dances and songs without elephant-foot drums. In daily life, if you step into the thresholds of Dai families, hosts would propose toasts of wine and entertain you with hospitality and sincerity.

Shui people value wine so much that they often treat each other with meat and wine. Drinking wine is a must on important occasions such as weddings, funerals, festivals, visiting families and friends, commencements of hunting and cultivation, and building wells and houses. Not only men are good at drinking, women can also drink. When entertaining female guests, housewives often enthusiastically persuade them to drink.

Wa people have the custom of "showing courtesy with wine". Therefore, wine has become an essential item for important life events, such as receptions and negotiations, weddings and funerals, and building houses. All families prepare wine at home, so there is wine fragrance in Wa villages in all seasons. Wine not only plays a role in regulating and balancing the relationship between people, but also brings joy and interest to the life of Wa people. Most Wa women can make wine. Ever since ancient times, Wa women have been evaluated according to their competence in brewing sweet wine. Such competence is also something that Wa women are proud of. Usually, the hostess of a family would be respected if the wine brewed by her is the sweetest and well known far and near.

Brewing wine is the most basic life skill of Achang women. Whether they can make good wine has become an important criterion in evaluating girls' talents. Achang girls have been learning how to make rice wine and liquor since they were young. After the autumn harvest, singing love songs, young people would go up the mountain together to collect 18 kinds of herbs such as bitter grass to make distiller's yeast. During the twelfth month of the lunar year, every family makes fire to brew wine. They would keep the wine preciously in urns for guests or festivals in the coming year.

Wine is also found in the daily lives of ethnic minorities such as Hani, Achang, Mongolian, Pumi, Yao and Dulong. Wine represents their common hospitality, advocacy of sincerity, unity and friendship. Either ascending the bamboo houses of Dai families, walking into the Tuzhang houses of Yi families, the wooden houses of Pumi people, or entering the Mongolian yurts, the cottages of Bai people, the guests will be warmly entertained, and all ethnic groups will serve wine according to their traditional customs.

Wine as an auspicious symbol in new house

Ever since ancient times, Yi people have followed the custom of drinking wine twice when building new houses. The first time is when setting up a pillar. The second time is the "Opening the Gate" ceremony held when the gate of the new house is installed. Relatives and friends would bring gifts and the owner of the new house would hold a banquet to entertains guests.

On the night of the day when Nu people commence building a house, the host would entertain builders and express gratitude with toasts of wine. Builders would also bring wine to congratulate the family.

After the completion of the new house, Jingpo people will hold a ceremony of "Entering the New House". When entering a new house, the Jingpo family must first choose a good time to bring in the new fire, usually in the day time. Dong Sa (the priest of Jingpo religion) takes the lead to walk in the new house, new fire (torch) in hand. Then follow the host, with an iron triangle, an iron pot, a tube of rice wine and a tube of clear water, and the elderly hostess, with a basket of millet. A large-scale mass dance is held that night. Guests would congratulate the owners of the new house with the song of "He Xin Fang" and toasts of rice wine. In this ceremony, wine is essential for creating jubilant atmosphere. It serves as the only drink in the collective dance all night long. The fact that it is brought into the new house together with such daily necessities as fire, iron triangle, iron pot, water and millet demonstrates the status of wine in Jingpo people's hearts.

For Achang people, setting up pillars and installing beams for new houses are festive occasions. On that special day, the owner of the new house would prepare abundant wine for worshiping gods, serving house builders and entertaining relatives and guests.

When Dai people move into a new house, relatives would gather near the new house on time. A man with a new rice steamer on his back, a long knife on his shoulders, and an ignited dry cow dung cake in his hand, will lead the way to the door. Then, relatives and friends will come to the gate of the house with gifts one after another. After formal greeting, the host invites guests into the house and entertains them with fragrant tea and mellow wine. Having gone upstairs, the guests place the kindled cow dung cake into the fire pit to make a new fire and cook the first meal for the host on the new fire. It is said that the ignited cow dung cake symbolizes the eternal existence of the new house, and the life of the new house owner will be as exuberant as the new fire. The long knife on shoulders suggests that the host family can avoid evil, eliminate disasters, enjoy peace and health in all seasons. The gifts presented by relatives and friends have their special symbolic meanings. There are live chickens whose legs symbolize pillars. Dai people believe that the more pillars there are, the stronger the house is. There are betel nuts which symbolize sweet happy life. The more betel nuts Dai people chew, the sweeter their life would be. There are also white threads which can bind the owner's soul to the new house, so that the owner would live longer, the house would be strong, and there would be abundant grains and prosperous domestic animals. There are cooking utensils which express wishes that the host family would not worry about food and clothing in the future. On that day, the host entertains relatives and friends with wine and delicacies such as pork and chicken, while guests sing the song of He Xinfang for the host.

After the completion of the new house, Lisu people will hold a grand ceremony which is called "Entering the New House". The most important event in the ceremony is igniting fire in the new fire pit by virtuous elders. The vigorous new fire indicates that the host's family will be prosperous and everything will go well. Food cooked on the new fire would be offered to ancestors. Relatives, friends and villagers will also come to congratulate with homemade wine and gifts. Gift givers must come in pairs. If there are few people in the family, the new-born baby could also be counted as a gift-giver. People gather in the new house, enjoying wine, singing and dancing. The song "Building a House" is indispensable. The hostess and a guest sing in antiphonal style. Others join the chorus while dancing as the house-warming party reaches its climax. Finally, the hostess will sing:

Gratitude to friends from the east of our village,

Gratitude to friends from the west of our village,

My sincere gratitude to you all.

Now I can sleep soundly,

Now I can sit peacefully,

Why do I say that?

Because I have a three-room house now.

Appreciate work done by friends from the east of our village,

Appreciate work done by friends form the west of our village.

I have no delicious food to offer, no fragrant wine to provide,

But you still came to help build my house.

From now on, I can do my work at ease,

From now on, I can go to work without worry.

Next year, on the same day,

Next year, in the same month,

Please come for meal again,

Please come for sorghum wine again.

There are two important things to do when Lahu people celebrate the completion of their new house. One is to set up the sacrificial altar for the house god, burn incense and candles, offer rice and water, and pray that the house god blesses the family with auspiciousness and peace so that the family would be better off. The other is to make a fire pit next to the center pillar of the house, place a triangular iron pot stand in the fire pit, burn incense and pray near the fire pit. Once the fire pit and the triangular iron stand are placed, no one should move them. Not any outsider is allowed to touch them, otherwise such things will be considered adversities. After worshiping the house god and setting up the fire pit, the owner of the new house will prepare a sumptuous banquet. The villagers who come to help build the house are invited. At the banquet, elder people chant the song "He Xin Fang". All guests sing and dance to the SanXuan (a three-stringed plucked instrument) and the LuSheng in front of the new house, extolling the completion of the new house.

After the new house is built, Hani people will hold the ceremony of "Yong Dada"(similar to house warming part). First of all, a respected elder in the village takes the lead to walk into the new building with eggs and some glutinous rice. The elder is followed by a group of young people carrying a triangular iron pot stand, pots, bowls and chopsticks. After everyone has entered the building, the elder puts the triangular iron stand on the fire pit, makes a fire, and invites the god of warmth and light into the new house. From then on, the fire in the fire pit will burn forever, which symbolizes the family's life, fortune, and abundance. After the ignition, the eggs are peeled and mixed with glutinous rice for everyone present to have a taste. After the ceremony, the owner of the house prepares a banquet to thank relatives, friends and villagers who participated in the construction of the new house. The owner would also invite "Ya Xi" (singers) to sing hymns. Hani people believe that the future of the host family's life is closely related to the lively atmosphere of the ceremony. Therefore, the new house owner would make the banquet and the celebration as sumptuous, jubilant and joyous as possible, expecting that their future life in the new house would be as pleasant and abundant as the ceremony.

Zhuang people must choose auspicious days for moving. The most important thing for the host family is to invite the master of the village to recite sutras and move the ancestral tablets, shrines, and incense burners to the new house. Relatives and friends gather to celebrate, helping the host family kill pigs and offer sacrifice to their ancestors. The enthusiastic atmosphere is full of expectations for a new life. When relatives and friends are at table, everyone drinks wine and congratulates the host family with best wishes for prosperous, peaceful, and harmonious life. Guests will also sing "He Xin Ju Song":

I haven't been to this part of the village for long,

Where stone slabs are tightly connected,

And the road is as flat as slate.

Your new house is on the side of the road,

Sitting north, facing south,

With plenty of shade and cool.

The rocky mountains on both sides are like lions,

The rocky mountains in the front are like

two phoenixes flying in the morning sun.

The first flight of steps

Are like a grand front terrace which welcomes

the blessings of spring, summer, autumn and winter,

And receives fortunes from east, west, south and north.

……

Congratulations to the hall, congratulations to the beam.

The red central beam is eight Zhang long (1 Zhang=3.33meters),

With Fu, Lu, Shou (luck, fortune, longevity) carved on it,

And dragons and phoenixes in the morning sun.

The new handrail is high and wide,

Wishing you prosperity for generations to come,

Wishing you industrious children and prosperous generations.

Peace to your descendants and longevity to you all.

Wine as a cooking condiment in cuisine

In the rich and colorful ethnic minority food culture, wine is an important condiment. It has been widely used in cuisine by all ethnic groups in various ways. When wine is added to common dishes such as fish, stew, stir-fried or cold dishes, the color and fragrance of the dishes become more attractive. One unique ethnic cuisine is the use of distilled wine as water to cook meat.

The wine-boiled chicken of Pumi, Lisu, Nu, and Dulong people in the Nujiang Canyon in western Yunnan is a unique traditional ethnic delicacy, which is called "drunken chicken". Nu people and Dulong people call such dish "XieLa", meaning wine-stewed meet. The cooking method of Xiela is as follows. Cut chicken into small pieces. Fry the chicken until it turns golden. Add 2 catties of distilled wine and stew the chicken for half an hour. The drunken chicken cooked in this way has a mixed taste of meat and wine. It is not only a traditional delicacy in the Nu River Valley but also a nutriment for the elderly and puerperal women.

Hani people and Dai people living in the river valley of southern Yunnan have a custom of cooking "drunken fish". They put the freshly caught live carp into rice wine. Let it swim in the wine until it is drunk, and then stew it. The taste is enticing. There are also drunken ducks and drunken geese. Shrimps could be served raw after dipped in liquor or cooked with rice wine.

Wine, used as a condiment, could add tenderness to fried meat and vegetables. When ethnic minorities stir-fry or stew meat, they would soak the whole piece of fresh meat in wine to reduce its fishy smell and retain its natural color and flavor. Wine added to fried vegetables could also retain the color and flavor of vegetables. Vegetables cooked in this way are appetizing and easy to digest. For example, when Dulong people and Nu people cook poached eggs, they would add a little distilled wine and simmer the egg for a while. Thus the poached egg becomes delicious with wine flavor. Nu people often mix eggs with freshly brewed hot wine. Such mixture is nutritious and delicious.

All ethnic groups have traditional customs of pickling sauerkraut, chili, milk curd, and other pickles with distilled wine as a subsidiary ingredient. When pickling, mix the main ingredients with Amomum tsao-ko powder, star anise powder, Chinese Prickly Ash, ginger slices and chili powder, and sprinkle a little distilled wine before storing the pickles in crocks. While playing a role in sterilization and antisepsis, wine can help retain the flavor and freshness of the pickled products. After being sealed and stored for a period of time, the pickles will taste delicious. Although each kind of pickle has its own characteristics, they all exude the aroma of distilled wine.

Sour fish is a traditional delicacy of the Dai people living along the Red River in southern Yunnan. Its unique flavor is closely related to the origin of the fish. During the first month of the lunar calendar, Dai people go to the hillside to mow thatch, tie them into fan-shaped grass nets, put them in the river, and wait for the carp to lay eggs on the grass nets. They bring the grass nets with fish eggs back to their own fish pond for the fry to hatch there in more than ten days. After the early rice is planted, the fry are transplanted into the paddy field. When the harvest season comes, carps have grown to 6-10 inches long, ready for cooking. Open the belly of the fish and take out the internal organs, sprinkle the fish with salt, Amomum Tsao-ko powder, and Chinese Prickly Ash, and fill its belly with glutinous rice wine. Then store and seal the fish in a crock for more than half a year. When the fish flesh becomes light yellow and the fish bones are soft, it is ready to eat. It can be eaten raw. If deep-fried, it is crispy outside and tender inside, with a mixed fragrance of fish and wine.

Wine as a potion for health and longevity

The medical effects of wine are manifested in medicated wine, but wine itself also has certain health care functions. An appropriate amount of wine can speed up blood circulation, promote metabolism, enhance immunity, stimulate saliva secretion, and improve digestion, which is beneficial to health. For both ordinary people and the elderly whose organic functions are gradually declining, wine is good for health and longevity. Therefore, ethnic minorities also use wine for health care.

For example, Zhuang people believe that wine can relieve fatigue, soothe the muscles and invigorate blood vessels. Drinking wine after having walked or worked long in mountains, they would feel comfortable and relaxed. They dissolve the bile of chickens, ducks, pigs, and cows in wine. It is said that such mixture is refreshing and can improve eyesight. Zhuang people also make precious medicated wine such as gecko wine, three-snake wine, and tiger bone wine. Gecko wine nourishes lungs and kidneys. It is a good aphrodisiac. Three-snake wine and tiger bone wine have special therapeutic effects on rheumatism, bruises, and lumbar muscle strain. In the long-term practice of brewing and drinking, Achang people have accumulated rich experience. They use wine for medical care and have developed a variety of medicated wines for the treatment of various diseases such as bruise, stomachache, algomenorrhea, rheumatism, etc. They also use wine for warmth, nourishment, disease prevention and so on. Mongolian horse milk wine contains a variety of amino acids and vitamin C, which can relieve abdominal distension, intestinal cramps, and diarrhea caused by the lack of lactase. The horse milk wine has the functions of nourishing body, warding off the cold, invigorating blood vessels, nourishing kidneys, and promoting digestion. The glutinous rice wine of Bai people is specially made for pregnant women because it is a nourishing lactagogue.

Yao people believe that liquor can dispel dampness and ward off cold, relieve fatigue, invigorate stomach, and strengthen body. Wa people admire wine and believe that wine is the noblest and most holy item. It can help Wa people ward off cold, exorcize evil spirits, and detoxify. It can also nourish body, strengthen courage, and refresh spirit. Therefore, wine is widely used by Wa people in all aspects of production and life. Learning from their life experience, Wa people know that wine has a good detoxification effect. For example, if poisoned by wild grass or fruit, one can get better soon after drinking a few sips of distilled wine. If injured, wash the wound with distilled wine so that the wound will heal quickly rather than swell.

In addition to concocting medicines, curing diseases, and swallowing medicines with liquor, ethnic minorities also add the roots, fruits, and stems of plants, animal bones, gallbladders, eggs, etc. into their self-brewed distilled wine to make various types of medicated wine. Both adding herbs to wine and taking herb medicine with wine are for disease prevention and treatment, and longevity.

Section 2   Wine and Festival Celebration

Wine is closely related to festivals of various ethnic groups. People toast to celebrate good harvest, commemorate the achievements of ancestors, express their feelings, drive away evil spirits, eliminate filth, boost morale and pray for blessings. Observing the festival customs of various ethnic groups, we could say that there is no festival without wine. The various customs about the use of wine in festivals can reflect the characteristics of ethnic festival culture. These customs have become a window for people to know and understand ethnic minorities.

Wine as a magic power for exorcising evil spirits

Wine , an ordinary drink discovered by the ancestors in nature, has special magical power in the minds of ethnic minorities. It satisfies the religious needs such as exorcising evil spirits, eliminating filth, worshiping gods and ancestors. Many ethnic festivals originated from religious activities. For example, the popular Torch Festival among the Yi language ethnic groups originated from their ancestors' worship of fire, while the Dai's water-splashing festival started from the water worship common in the Dai society. Bai's Raoshanling, Naxi's Sky-worshiping Festival, Jingpo's MunaoZongge, Zhuang's Gumiji Festival, and Lahu's Koumuzha, all reflect the the widespread ancestor worship and nature worship among minorities. In these festivals, wine is fully deified. Moving from the dining table to the altar, wine has evolved from the form of material culture to the form of conceptual culture.

The most grand traditional festival "Munao Zongge" in the Jingpo society brings the supernatural power of wine into full play. Munao Zongge is also called "Zongge", which means "happy gathering, singing and dancing". Its main purpose is to eliminate calamity, exorcize evil spirits, pray for peace, and celebrate good harvest and tribal victory. "Munao" is Jingpo language, and "Zongge" is a literal translation from the language of Zaiwa people, meaning everyone dances together. Munao Zongge Festival is the carnival of Jingpo people. It has the reputation of "Dance of Heaven" and "Carnival Dance of Ten Thousands of People". In the settlements of the Jingpo ethnic group in Dehong Prefecture, Munao Zongge is held in every village around the 15th of the first lunar month.

Before the ceremony comes, all households in Jingpo villages make rice wine. Their gardens are filled with pleasant traditional wine songs as the rich and intoxicating aroma of wine rises above trees. During the festival, Jingpo people gather together happily with bamboo tubes full of self-brewed wine on their backs. Wine is not only the most eye-catching drink, but also a gift pleasing to gods and ghosts. Therefore, a large amount of wine is used to worship gods and entertain ghosts during the Munao Zongge Festival. Whenever there is Munao Zongge, the most important preparation activity for Jingpo people is to prepare wine for the ceremony. Jingpo people have no doubts about the sacred power of wine, but earnest pious emotions. Therefore, there is also one Sijiu (the person in charge of wine) on important occasions. Sijiu must be appointed by a Jingpo wizard and highly respected. Its status is generally second only to the host of the Munao Zongge Festival.

Munao Zongge has a strong cultural and entertaining color, and has become a grand gathering for entertainment and trade. The traditional sacrificial dance Munao dance is the real climax of the festival. After a series of activities such as offering symbolic sacrifices to good ghosts and expelling evil spirits, the Jingpo people who participated in the Munao Zongge celebration would dance under the leadership of "Naoshuang" (lead dancer). Thousands of people danced to the same drum, presenting a spectacular scene. Once started, the dance will last for more than ten hours. When people are hungry, they drink wine to keep off their hunger; when they are thirsty, they drink wine to soothe their throats; when they are tired, they drink wine to refresh themselves. Accompanied by the typical Jingpo music, people enjoy the auspiciousness and joy of life in their dance. Wine, elevates Jingpo villages into a happy and peaceful paradise.

The wine custom followed by Nu people during the Chaoshan Festival is another example of the sanctification of wine as a material form. The Chaoshan Festival is also called the Flower Festival because it is held on March 15th of the lunar calendar, when the spring flowers are blooming. It is the most solemn traditional festival of the Nu people in

Gongshan Dulong and Nu Autonomous County, Yunnan Province. Legend has it that a long time ago, there was a girl named A Rong in a Nu village. She was not only beautiful, but also smart. One day, when she was weaving at home, she saw a spider weaving a web under the eaves. She was inspired and invented the sliding cable. Together with the villagers, she cut the golden bamboo, wove it into a sliding cable over the surging Nu River to eliminate the difficulties of crossing the river for villagers. A Rong was hailed as a fairy because of her invention. When the slave owner heard about it, he sent a matchmaker to propose, but A Rong refused. The slave owner became angry and sent his servants to abduct A Rong for marriage. A Rong hid in a stalactite cave deep in the mountains after learning about it. The angry slave owner instructed his servants to set the mountain on fire. Unfortunately, A Rong died in the fire. To commemorate her, Nu people designated the day of A Rong's death as the "Flower Festival."

The major activity of the Nu Nationality Flower Festival in Gongshan County is that people bring their self-brewed rice wine to the caves in the mountains to offer sacrifices to "fairy milk". The so-called "fairy milk" is the water dripping from stalactites regarded as fairies. People "invite" the fairy milk home reverently, dance three times devoutly around the pillars in their houses, then mix the fairy milk into the self-brewed wine. Men, women, and children all drink a bowl. After drinking this bowl of wine with a taste of "fairy", they start a feast, singing and dancing. Nu people believe that after their drinking this bowl of wine, the fairy will bring luck and peace to their families and villages.

On the Flower Festival, Nu people get up early in the morning, put on festive costumes, bring sacrifices, wine and various foods to the "fairy caves" near their villages. The flowers on both sides of the Nu River are in full bloom. They gather bunches of flowers and place them around the fairy cave for the fairy A Rong. They also hold a sacrifice rite in front of the cave to pray for auspiciousness and good harvests. After that, they sit with families and friends on the hillside, lay the prepared food on the ground covered with pine needles, and start eating and drinking. They sing and dance while eating and drinking. The hillside is full of joyous festive atmosphere. In the evening, young men and women light a bonfire. Around the bonfire, young people sing love songs and dance happily all night long.

Torch Festival is not only a festival of Yi people, but also an important traditional festival of the ethnic groups of Yi language such as Naxi, Jinuo, Lahu, Hani, and Lisu, as well as Bai people. But the days for celebration are different. Yi, Naxi, and Jinuo observe the Torch Festival on June 24 of the lunar calendar, while Bai's Torch Festival is on June 25, and Lahu's on June 20. The festival lasts two to three days.

The Bai torch festival originated from the legend of "Burning Songming Tower". The legendary protagonist became the deity of Bai people in a disaster caused by wine. Therefore, during the Torch Festival, wine must firstly be sacrificed to worship the god of fire, then worship Benzu (sanctified ancestors or heroes) before being enjoyed by Bai people. Before lighting the torches, offering wine to the god of fire and various related gods is an important matter. In order to ensure that the torches have raging flames, Bai people would prepare liquor, pine resin, and cooking oil. When walk into the dim places of a house with a torch in hand, they spray liquor and sprinkle pine resin on the torch to elicit raging flames so as to expel the evil spirits and filthy diseases hidden there. Such sacrificial rituals extend the material properties of wine from real life to the spirit world while sanctifying the function of wine.

Wine occupies an important position in Yi's festivals. During the Torch Festival, young Yi girls will bring freshly brewed corn wine and beautiful, exquisite wine utensils to the road side, and set a wine array called "girl wine". They offer wine as a gift to elders, friends, relatives, and guests who participate in the festival activities. After wrestling, horse racing, bullfighting and other competitions, the girls will propose toasts to the victorious wrestlers and Yi warriors and sing toast songs to express their appreciation and admiration. In some places, the young men and women of the Yi ethnic group "drink festival wine". Before the Torch Festival comes, young Yi are busy preparing for the "festival wine". Young ladies buy cloth to sew new clothes, new pants, new cloth bags, and store wine in secret places. Young men save money for felts and capes, floral cloths, yellow umbrellas, and gifts such as silver collar buckles, earrings, bracelets, Jew's harps and candies. On the second day of the Torch Festival, young Yi who are in love with each other will find a suitable place and sit together in groups. Young ladies would first hand the wine to their respective lovers. After complimenting the wine, young men would place the wine and stewed piglet, roasted whole chicken, fried oatmeal flour and other dishes in front of the young ladies. They would put on the new clothes presented by their lovers and let their peers appreciate. Then, young men would present the prepared gifts to their respective sweethearts. At last, all of them would sit together and enjoy the delicious food while chatting.

The the origin of the Dai Water-Splashing Festival directly introduces wine into the world of gods and demons. According to the legend, Dai people used wine to conquer demons and exorcize evils. A long time ago, there was a fierce demon king in the Dai ethnic area. The king was so lewd that wind and rain became out-of-order and crops could not survive. He abducted seven Dai girls to be his wives. After the seven sisters used a trick to find out the deadly weakness of the devil, they took turns to get him drunk with fine wine, and quietly pulled a wisp of hair from his head. When they tied the hair to the devil's neck, his head fell off. But as soon as his head hit the ground, fire broke out. Seeing that the beautiful homeland of Dai people would be harmed, in desperation, the eldest sister picked up the devil's head from the ground, and the fire went out all at once. So the seven sisters took turns holding the devil's head in their arms. Whenever there was a rotation, the sisters poured clean water on each other to wash away the filth on their bodies. In order to express their respect for the seven sisters, Dai people celebrate the Water Splashing Festival every year on the days when the seven sisters killed the evil king.

Wine as an emotional bond on festive occasions

In ethnic festivals, wine customs are beneficial for unity of ethnics, exchange of ideas, expression of emotions, and the festive atmosphere. Drinking is for cheerful mood and warm atmosphere, and is the main theme of ethnic minority festivals.

The Miao minorities love wine. After work, they drink a bowl to relieve fatigue and restore their strength. "Steamed ground corn mixed with liquor"is the most favorite diet of Miao people. In leisure time, they would enjoy wine with confidants either among flowers or in the moonlight. As inspired by wine, they revel in singing and dancing. In festive seasons, wine becomes the best company to them. "Cai Huashan" is a grand traditional festival of Miao people, popular in Miao villages in northeastern and southern Yunnan. The time to celebrate the Cai Huashan Festival varies from place to place, but most of them are at the beginning of the first month of the lunar calendar when spring comes and mountains are full of flowers. The locations for the celebration are usually the clearings beside the villages. It is said that the Miao leader Chi You and the Han leader Huangdi had a battle. After Chi You was defeated, the scattered tribe members were ordered to gather once a year at the beginning of the year. The later generations designated this time as the Cai Huashan Festival. The Cai Huashan Festival was originally intended to worship Chi You, the ancestor of the Miao nationality. Nowadays, there are more activities such as worshiping and climbing the flower pole, singing and dancing, bullfighting, martial arts performances, etc. Either before or during the festival, all activities are closely related to wine, revealing the cultural function of wine at Cai Huanshan.

On the day for Cai Huashan celebration, Miao people will cut a tall tree, prune branches to the crown, and plant it in the middle of the celebration venue. Miao call this tree the "flower pole". According to the legend, it evolved from ChiYou's battle flag. They fill the gourd jugs with their home-brewed wine, hang the jugs high on the crown of the flower pole, or placed them at the foot of the pole. Red, yellow, blue and white ribbons are tied on the pole for revel atmosphere. Young people swarm to the flower pole from the villages around, men in short blouses, with cyan headbands and cloth belts on their waists, women in costumes with embroidered patterns and jewelries such as silver earrings, bracelets, rings, necklaces, etc. They sing and dance around the flower pole with the beats of the Lusheng, SuoNa, HuQin and other ethnic instruments. While singing and dancing, if a young man finds the girl he likes, he quickly removes the umbrella carried on his waist and holds it up for the girl. If the girl does not like the young man, she would immediately go around to hide among other girls. If both of them are delighted in each other, they would express their heartfelt feelings under the umbrella and learn more about each by singing antiphonal songs. In addition to singing and dancing, there are intense competitions such as horse racing, archery, shooting, climbing the flower pole, and so on. Among them, bullfight is the most popular and the fiercest one, yet with jubilant atmosphere. Take the bullfight in Qiubei and Guangnan for instance. The organizer must prepare mellow wine before the event. As a reward, the host will propose three toasts to the owner of the champion bull in public. As a punishment, the host will also pour three glasses of wine and let the owner of the defeated bull drink in public. This judging principle, namely three cups of wine for winning and three cups of wine for losing, reflects the entertaining essence of the Cai Huashan Festival and Miao people's characters of kindness and generosity.

The October New Year is the most splendid festival of Hani people. It is called "Zhalert" in Hani language. Every household in Hani villages would make glutinous rice cakes and Menguo liquor (distilled wine) before the festival arrives. During the Zalert festival, the most nationally and culturally typical celebration is the Long Street Banquet, called "Zigbaduo" by the Hani. Few banquets of ethnic minorities could attain its magnificence, revelry, and fervency. The Hani language "Zibaduo" means taking turns to invite each other to drink. On the afternoon of the Zibaduo celebration, every household in the host village should give full play to its cooking expertise, prepare a table of wine and delicacies, carry them to the street or somewhere in the village as agreed beforehand. When all wine and delicacies are placed on the long line of tables, all villagers, young and old, take their seats and the annual long street feast officially begins. This kind of national banquet is a manifestation of the "one person obtains, everyone shares" distribution system of Hani ancestors. It also strengthens the cohesion of the villagers. At the Zibaduo banquet, the younger generation must pour wine for everyone, and the elders must raise their glasses first. While enjoying the homemade Menguo liquor, elderly people sing the ancient Hani folk songs. Young people would toast to the old people one after another. When elders accept the respect of the younger generation by drinking up the wine in their glasses, all would cheer in unison. People eat along the long line of tables, from the "dragon head" table to the "dragon tail" table, wishing each other the best, a prosperous family, peace and happiness. Their laughter echoes in the old streets and the verdant forests.

The New Year of the Yi nationality, called "Kushi" in Yi language, is observed on the lucky days during the first ten-day period of the lunar October, lasting for 3-6 days. It has become an important traditional festival for Yi people to remove the old and welcome the new, commemorate the ancestors, and pray for good weather, abundant grains, and good luck in the coming year. Therefore, it is a grand festival for Yi people. The celebrations of the Yi living in the Liang mountains are the most distinctive.

In the early morning of the first day of the Yi New Year Festival, young people set off firecrackers. Women sing auspicious songs while pounding glutinous rice cakes, roasting buckwheat cakes, boiling eggs, and cooking rice. Men kill pigs and sheep, prepare food such as tuo tuo meat, and welcome their ancestors home for the festival. After the New Year's Eve dinner, middle-aged men would gather in groups to bring their best wishes to every household. They hail when they are at the door to inform the host about the arrival of guests. After entering the courtyard, they hail again before stepping into the house. The host would take out wine to entertain everyone. After drinking, everyone hails happily to express their gratitude. If the host brings out good wine again, people will hail merrily, praising the host for his generosity. Women stay at home to entertain visiting relatives, friends and guests. In the evening, the family sits around the fire pit for their ancestors to count the number of family members. Such tradition of siting with ancestors is called Peiye (accompanying ancestors through the night). The oldest person in the family will sacrifice buckwheat cakes and eggs to worship the ancestors. Then, family members drink and eat grilled meat, talking and singing, fully immersed in joy and happiness.

On the third day of their New Year Festival, Yi people will see their ancestors off by offering sacrifices. Before dawn, each family fries the rice served on the sacred table, recooks the meat, and makes three or seven buckwheat cakes. They sacrifice the fried rice, the recooked meat, the newly-made buckwheat cakes, together with two packages of noodles, and a bag of tobacco leaves so that their ancestors could eat before heading back to the heaven. When roosters crow for the first time, all the family members sit by the fire pit to have a bit of the sacrifice, and wish each other all the best.

During Yi people's New Year Festival, young men wear black shirts, a head wrapper made of black cloth, and a yellow or red ear ring on the right ear. Girls wear garments, multi-pleated long skirts, with multi-layered colorful cloth on the hem, and a black kerchief on the head. People gather to sing and dance with the Jew's harp, Yueqin, Huqin and Lusheng. They sing antiphonal songs, dance, swing, or participate in horse racing, archery, wrestling and other activities. Yi people, though living ten miles away, would bring their families, old and young, to join in the festive activities.

The first day of the first month of the Tibetan calendar is the Tibetan New Year's Day. Before the New Year, the courtyard is cleaned on the 29th of December. After that, a large fire is lit in the courtyard and the whole valley is filled with smoke. The pictures of the propitious umbrella, goldfish, vase, lotus, conch, auspicious knot, Choggi Gyalshan, and

Chakra are also drawn on the doors and beams of the house with alkali or tsampa (roasted barley flour). The offerings for worshiping the gods on the first day of the first lunar month are prepared, including fried food, "Chema" (a square grain bucket filled with barley and butter flower, symbolizing auspiciousness), wine, tea, and silverweed cinquefoil roots.

Before dawn on the first day of the New Year, each family will go to the river to fetch the first bucket of water, called auspicious water. The family's daughter-in-law or eldest daughter is responsible for making the wine. When the cock crows, she will bring the families the "Guan Dian", namely boiled hot barley wine with brown sugar and milk dregs. When dawn is approaching, beautifully dressed young Tibetans, carry wind flags, divine incense, Chema, tsampa, and barley wine to mountains. They burn incense on the top of the mountain, plant wind flags, and worship the mountain gods. They sprinkle some tsampa on the ground while praying "good harvest for an abundant year". When all the worshiping rites are done, they race down the mountains. The first one to reach the foot of the mountain, will be rewarded with three glasses of wine, and the penalty for the last one will be four glasses of wine.

The morning of the new year day is also called "the luster morning". Tibetans would not only anoint children's heads with oil, but also mix the soot and oil to polish the horns of the livestock, and feed distiller's grains to the livestock. The wealthy families also feed livestock with newly brewed wine, glue butter flowers on the tips of the horns, and replace the livestock's old earrings and old collars with new ones. At breakfast, the mother puts the barley wine in the middle of the seats, and glues three butter flowers on the side of the wine jug. The eldest daughter holds the wine jug, starting with her parents, taking turns to propose toast to family members. Then the mother serves the oatmeal porridge, which contains cheese, silverweed cinquefoil roots, meat and so on. After drinking the oatmeal porridge, they start to drink the after-meal wine which is usually poured into large wooden bowls and horns. No drop of wine is to be left in the bowl, otherwise you must drink one more bowl.

In the following three to five days, friends and relatives visit each other with best wishes. The guests enter the door with greetings of "Tashidler", the host immediately greets the guests with "Tashidler" in return. Some hosts even present Hada to guests. Together, they enter the room and sit on the new mats. When the host brings tsampa, the guests sprinkle some tsampa into the air to worship gods in heaven and on earth, then put a little bit in their mouths. The host will bring out a jug of barley wine and invite the guests to drink. In order to show respect to the owner, each bowl of wine must be finished with three sips. If one can't finish the wine, the hospitable host will entrust relatives and friends to sing a toasting song. When the singing is over, the guest must have drunk up the wine.

Wine is also essential in other Tibetan festivals. The Archery Festival is a traditional festival of the Tibetans in Deqin, Yunnan. It is in the fourth lunar month. Before the festival, adult males gather to discuss matters related to the festival, and elect a host who will be responsible for preparing arrows and wine. Each participant will hand in one arrow and two or three catties of barley, which are then brewed into wine. All men, regardless of age, participate in archery. Those who are unable to shoot can choose a substitute. At the beginning of the festival, a ceremony is held first, then all participants are divided into two groups. Having drunk the wine, they enter the arena. Each person can shoot two arrows and the one who scores the highest wins. There are three to five rounds of competition every day. At night, the women go to the shooting range to toast, bless the shooters, and light bonfire for singing and dancing. Tibetans love the nature so much that, adapting to the plateau climate, environment and living conditions, they have formed a unique custom, that is, "walking around the forest". Every year from May 1st to 15th of the Tibetan calendar, Tibetans walk out of the courtyard to the shady woods, set up tents, enjoy the gifts of nature, sing and dance while drinking barley wine. The Xuedun Festival is one of the most famous traditional festivals in Tibet, lasting from June 29th to July 1st of the Tibetan calendar. In Tibetan language, "Xue" means yogurt, and "dun" means feasting. Literally explained, "Xuedun" is a holiday for eating yogurt. Because there are Tibetan opera performances during the Xuedun Festival, it is also called the "Tibetan Opera Festival". During the festival, Tibetans, young and old, set up colorful tents under the shade of the trees and display festive foods such as barley wine and delicacies on carpet. They chat while eating, and sing while dancing. When guests come to the tent, the host would toast to the guests and sing wine songs of different tunes in order to persuade them to drink. Then, they toast to each other in the warm and enthusiastic atmosphere.

The only traditional festival of Dulong people is the New Year Festival. The 29th of the twelfth lunar month is the New Year's Eve, and the 30th is the first day of the new year. Dulong people call this festival "Kaquewa". During the Dulong New Year Festival, each clan and tribe must hunt wild animals collectively and distribute the prey to each household. On the New Year's Eve, relatives and friends flocked in. The host and the guests would drink Tongxin Wine, hugging each other, keeping faces close so that they could drink from the same bowl. In the early morning of the new year day, when the first glimmer of dawn appears, gongs sound around the village for the coming of the new year. After breakfast, with the ring of the gongs, villagers gather in the clearing near the village to celebrate according to their ancient and simple customs. People, regardless of age, gender, and family, hold hands and dance traditional Dulong dances. The elders bring delicacies contained in exquisitely woven Dulong rattan utensils and share food with everyone in their traditional way. Then, singing, cheering, and the sound of dance steps intertwine merrily, lasting for days as Dulong people take turns to organize such festive gathering. The whole Dulong village is diffused with a warm, cheerful, and peaceful atmosphere.

During the "Kaquewa" period, the most important activity is to slaughter the bull and sacrifice it to the sky. The bull-slaughtering sacrificial rite is presided over by the wizard Namusa. A venerable clan chief or the wizard leads a sturdy bull to the center of the village square and fasten it to a strong pole. Women hang beads and other ornaments on its horns. A chosen beautiful girl wearing a colorful dulong blanket puts a dulong blanket on the back of the bull. When the other sacrifices are exhibited, the chief of the sacrificial ceremony lights some pine tree branches and prays to Gemeng for peace and propitious life for Dulong people and their livestock. Then the wizard pierces the bull dead through its armpit with a sharp bamboo spear. Immediately afterwards, the wizard carries the bull's head on his back and leads the crowd to dance around the sacrificial bull in a circle. They play gongs and wield swords and spears while dancing. Then beef is cooked and shared, which brings the New Year ceremony to its climax. Everyone drinks wine and eats meat, singing and dancing to celebrate the coming of the new year and pray for a good harvest and prosperity for people and livestock. The banks of the Dulong River become a sea of joy. In the end, all the people who have participated in the bull-slaughtering ceremony receive an equal share of beef.

The most ceremonious festival of Zhuang people is the Spring Festival which lasts from 1st to 15th of the first lunar month. In addition to sewing new clothes, Zhuang people would make sweet rice wine before the festival, and predict their fortune according to the quality of the wine. Zhuang people in Qiubei county would offer a meal to their ancestors in the early morning of the New Year's Day. Baihu (the chief of a tribe) takes the lead. Before offering, each household sends a bowl of rice, three or five bowls of vegetables, and a pot of distilled wine to the head of their clan who will send the same amount of food and wine to Baihu. On that day, Baihu designates Bumo to offer sacrifices to ancestors. After the sacrificial rite, Bumo beats the gong to order all households to release livestock and poultry. At noon, Bumo beats the gong again, and the villagers gather at Baihu's house for a banquet. Food at the banquet has been prepared by the heads of each clan who, as representatives, send a pot of liquor and a piece of meat to Baihu so that all the villagers could have the Spring Festival banquet together. On the night of the New Year's Day, the seniors of each household bring pig hearts and tofu to the head of their clan for a gathering. The clan head also offers pig heart. The chef of the clan stews the pig hearts and tofu. When it is ready, all people take a seat according to their seniority. While enjoying distilled wine, stewed pig heart and tofu, they listen to the clan history told by the clan head, praise virtuous persons and their good deeds, and reach an agreement about public welfare matters. On the second day of the Spring Festival, the clan head will hold "Geng Lao Xing" (a rice wine banquet). When the clan head blows the bronze horn, clansmen would gather at his house. He would pour a bowl of sweet glutinous rice wine for each person. The wine is to be sucked with a reed pipe rather than drunk with a spoon or from a bowl. As they are drinking, clan issues will be discussed while disputes among clansmen will be mediated by the clan head.

Lahu people have the custom of drinking wine and tying red threads the wrist during New Year celebrations. On the morning of the second day of the new year, there will be a grand ceremony for New Year greetings. Everyone wants to give New Year greetings to their parents and elders, and all villagers will offer New Year greetings to the village chief, blacksmiths, and Moba. People usually bring a pair of glutinous rice cakes and a bottle of wine, kneel in front of the elders, kowtow, and then respectfully pour wine for the elders.

Having drunk the wine joyously, the elders tie red threads to the wrists of young people while singing songs of blessings. After that, the village chief leads all the villagers, who hold sugarcane in their hands, to the neighboring villages for collective New Year greetings. They would bring gifts such as wine, meat, and glutinous rice cakes. All the people of the hosting village, regardless of their ages, would rush to the village square to welcome their guests. Then hosts and guests dance in a circle to the tunes of the Lusheng. When feeling tired, they will drink a bowl of wine, rest for a while, and keep on dancing jubilantly until sunset. The Lusheng player and the elders would return to the house of their village chief. They sit around the fire pit, sipping liquor and tasting delicious meat, and keep singing until dawn. On the third day of the new year, some people would bring some rice, vegetables, a bottle of liquor, and some meat to the blacksmith's house for a potluck party. All men sit on the left side of the table while all women sit on the right, enjoying the potluck banquet. When the sun sets, all villagers gather once again. Each family sacrifices to their ancestors and burns the filth leftovers from the New Year's Eve before throwing them outside the village.

The Danu Festival is also called the Zhuzhu Festival and the Grand Mother's Day. "Danu" in the Yao language means elderly loving mother, so Danu is also a birthday celebration. Legend has it that in ancient times, there were two equally tall mountains. The one on the left was called "Blossi", mighty and majestic like a warrior standing upright; the one on the right was called "Miluoto", which looked like a girl in long dress. The two mountains moved closer to each other every year. After 999 years, they finally came together. On the 29th of the fifth lunar month, with an earth-shattering thunderbolt, the tall, handsome Blossi and the elegant, slim Milotuo came out from the cracks and got married. They had three daughters. Time flew quickly. Obeying her husband's instructions, gray-haired Milotuo sent her three daughters away to make a living. The oldest daughter carried a rake to the plains where she cultivated, had children, and multiplied into the Han nationality. The second daughter left with a load of books and formed the Zhuang nationality with her descendants. The third daughter took some millet and a hoe to cultivate the wasteland among mountains. Living and working in peace and contentment, she became the ancestor of the Yao nationality. The third daughter's hard work had brought fruitful harvests. Unfortunately, all the full-grained fruits were plundered by birds, beasts, and gophers in an instant. At this calamitous moment, Milotuo encouraged her distressed daughter, "Dark clouds in the sky and setbacks in life are inevitable. As strong wind can not twist majestic pine trees, difficulties will not scare the hard-working people. As long as they work arduously, life will be happy." She gave her third daughter a bronze drum and a cat. In the coming year, the crops grew even more gratifying. The third daughter beat the bronze drum to scare away birds and beasts, released the cat to eat up the gophers, and won a good harvest. In return for the nurturing of her mother, on May the 29th, the girl brought abundant gifts to celebrate her mother's birthday and her harvest. From then on, Yao people celebrate Milotuo's birthday as a harvest festival. Yao people living in Xishuangbanna celebrate this festival from May 26th to 29th of the lunar calendar. During the festival, every household will sacrifice 400 grams of Linen to commemorate Milotuo's birthday. Every family will prepare chicken and mutton, take out preserved wine, and have a meal together. Married daughters bring their children back to their parents' home to celebrate the festival. Everyone wears festive costumes for the jubilant celebration. Each village will set up a large stage for singing. As the bronze drum rings, people start singing and dancing on the stage and the village becomes bustling with joy and excitement.

The most important activity of the Danu Festival is the bronze drum dance. Five people are required for the bronze drum performance. Two persons play the bronze drums, one plays the bronze gong, one plays the leather drum, and one dances with bamboo hats. The bronze gong goes first, followed by the bronze drums and the leather drum rhythmically. The bronze drum has twelve sets of traditional playing methods which, from different angles, express farming, hunting and scenes of fighting with nature. The dancer with the bamboo hats intersperses among the above mentioned gong players and drummers, making humorous and ridiculous moves from time to time to arouse laughter among the audience. The sound of the drums are sonorous, the movements of the performers are simple and rustic, while the style is forceful. At night, lanterns and torches wind along the mountain paths, like a fire dragon swimming towards the gathering place. People dance monkey drum dance, rattan dance, animal hunting dance, mountain exploration dance, harvest dance, bull horn dance, Lusheng dance, colorful umbrella dance, etc. . After the dance, young people sing affectionate antiphonal songs, in the course of which some of them get engaged. The elderly collectively sing the Milotuo ode. Their singing is full of respect for Milotuo. In addition to the Milotuo ode, they also sing wine songs for toast. At the interval of each stanza, they toast and cheer together, not wanting to leave even when dawn breaks.

During the Dragon Boat Festival, young Pumi people put on festive costumes and head for the caves in mountains to worship--- Rao Yandong. It is a religious festival. People light oil lamps on the rock in the caves, burn a pine branch or a cypress branch to pray for a good harvest. They bring rice, meat, wine, honey, and fried noodles for a potluck party. Both adults and children must drink a few sips of medicated liquor soaked with Acorus calamus L. and Realgar, eat honey cakes, and pray for good health. After the potluck party, Pumi people would bathe in rivers or under waterfalls, singing and dancing. Some men, with their guns and dogs, would enjoy hunting and archery on their horseback to their heart's content.

Wine as a blessing for Farming

Drinking is often the actual daily need of various ethnic minorities. A bit of liquor before working in fields or rivers can increase body heat and reduce the negative effect of tough work on the body. Liquor drunk after a day's toil can relieve fatigue. For a long time, wine has been an indispensable drink in the daily life of Shui people and drinking has become their habit. Shui women are adroit at brewing, and men are good at drinking. Housewives will apologize if there is no rice wine when the guests come. In the busy farming seasons, they drink at almost every meal. According to "Guzhou Miscellaneous Records", "Miasma occurs at all times, especially in August and September. Poisoned by miasma, one would suffer from diseases such as temple pain, constant fever, dizziness and vomiting. Only by drinking till one starts sweating will the person be cured. A few cups of liquor in the morning and evening can get rid of the miasma." It seems that in order to eliminate the miasma and adapt to the environment, Shui people have been drinking for a long time. Their favorite glutinous rice wine is the best food for entertaining guests.

Opening the Seedling Gate with Spring Sowing Sacrifices   Spring ploughing and sowing are major events in agriculture. Many ethnic minorities will hold the "Opening the Seedling Gate" ceremony (the agricultural ritual for transplanting rice seedlings).

Miao people believe that Qiugong is the ancestor who taught people to exploit wasteland for farming, while Dangpo is the ancestor who cast the land. Therefore, every year, before transplanting rice seedlings, they offer sacrifices to Qiugong and Dangpo. When the seedlings are full grown and ready to be transplanted, Miao people will have the "Kai Yangmen" ritual. In Miao language, it is called "Gaixiejie", which means "begin to transplant seedlings". On that day, under the auspices of the village chief, fish, meat, wine, tea and rice are sacrificed to the ancestors. Then, the village chief plants in the public field of the village three clusters of seedlings which form a triangular shape. At the same time, three grass markers are set up beside the seedlings. Each household will also do the same things, worship their ancestors with chickens, ducks, fish, and meat, and then plant three clusters of seedlings in their respective fields. They pray that Qiugong and Dangpo will open the seedling gate of the heavenly granary so that they could start planting seedlings. The sacrificial offerings will be eaten up by the whole family. After that, the busy rice planting work officially begins.

Mid-March of the lunar calendar is the seedling season for Hani people living in the Ailao Mountains. Before spring plowing, each household must choose a lucky day to celebrate the Seedling Transplanting Festival, also called "Opening the Seedling Gate". On that day, the entire Hani village is immersed in a joyous atmosphere. Villagers welcome this seedling-transplanting day with simple traditional ceremony. Generally, Hani people must choose a lucky day for planting rice seedlings, such as the days of dragon, monkey, dog, and horse. On that day, every household sacrifices Tangyuan (glutinous rice balls) to heaven and earth. Early in the morning, the male seniors of each household should prepare cooked glutinous rice, chicken eggs or duck eggs, rice wine, and chopsticks made of artemisia twigs, and place them at the water outlet of their own paddy fields. Only after all the rituals are done can seedlings be transplanted. Three clusters of seedlings must be planted first. The first cluster represents human food, the second cluster represents animal food, and the third represents all the crops. Then they are covered with bamboo baskets which can not be moved until all the seedlings are transplanted. Following that is a quick lap of running around the fields, which indicates that this year's seedling transplanting can be successfully concluded as soon as possible. The purpose of the ceremony is to pray for elimination of natural disasters and plagues of insects, and blessings of good weather and good harvests.

According to the customs of Hani people, if one family plants seedlings and the whole village will help. The owner of the paddy fields brings mellow, sweet rice wine and passes it, bowl after bowl, to those who are planting seedlings. According to the custom, on that day, anyone who passes by the paddy field where the ceremony is performed, whether men, women, young or old, strangers or acquaintances, will be dragged down to the field by the owner to participate in this activity. Even if you don't know how to plant seedlings, your planting just a few of them would please the owner, because they are symbols of auspiciousness and happiness.

Every spring, Tibetans would choose auspicious days for spring plowing. On the plowing day, Tibetans, dressed in costumes, solemnly bring barley wine and Chema to their farmland. Each household send four persons, one holding Hada, one holding a wine bowl, one holding a jug of barley wine, and one in charge of offering Hada. They first anoint a bit of butter on the horns of the cattle that have just finished the first plow, tie a piece of Hada to the yoke and then offer a toast to the cattle. The rules for toasting are the same as proposing a toast to people: drinking up one bowl of barley wine with three sips. As they hold the nose rings, the cattle will open their mouths for the barley wine. Nowadays, Tibetans use walking tractors to cultivate land and sow seeds, but they still offer Hada and toast to the cattle first, then to the walking tractors and farmers. The commencement of spring plowing is not only a festival for cattle, but also a festival for people. In the position as appointed by the prophet, the host of the commencement ceremony symbolically plows the field and sows the first seed to announce the beginning of spring plowing. On that day, Tibetans will rest and entertain themselves. The next day they will formally start plowing and planting.

Celebrating harvests with Xingu Liquor   Before the autumn harvest, according to the traditional customs, the Hani people living in Yuanjiang, Yunnan, must hold a sumptuous ceremony for their harvest--- "Drinking Xingu Liquor". The so-called "Xingu Liquor" is made in this way. Each family brings back ears of grain that are about to mature from the field, and hang them upside down on the edge of a small bamboo fence on the top of the gable wall at the right rear of the hall, in order to ask the god of the family to protect the crops. Then, some of the grains are popped and soaked into liquor, together with the unpopped grains. Such Xingu Liquor will be enjoyed on a lucky day. Every household will cook a big cock and prepare a hearty meal. The elderly in the village are invited to drink first, then all family members would drink a few sips of the Xingu Liquor and satiate themselves.

The fifth day of the tenth lunar month is the Cattle-loving Festival of the Miao people living in Xinping, Yunnan. The ceremony includes three parts: rewarding and praising the cattle, celebration of harvest, and celebration of reunion. In the morning, adults are busy cooking chicken, duck, and other delicious food. Children go to the mountains to gather flowers and red maple leaves. Elders bathe the cattle, comb their leather carefully, then feed them with glutinous rice. When the cattle are satiated, women hang wreaths made of mountain flowers around the cattle's necks, and men decorate their horns with red maple leaves. Afterwards, the whole family bring the cattle to the cattle square for people to appreciate. On the cattle square, the elderly comment on whose farm cattle are well raised, healthy and robust. Young men exchange their plowing and raking skills with each other. Having praised the cattle and seen them disappear in distance, people go home to celebrate their annual reunion and harvest. While enjoying the delicacies, they discuss about how to protect the cattle for the winter. This eating and talking will go on until their cattle return home in the sun set.

During Chinese Spring Festival, the Pumi people in Lanping and Lijiang counties have the custom of toasting to passersby and feeding cattle with wine. On the morning of the first day of the lunar new year, Pumi people prepare food at home, then bring some wine to the roadside of the village to wait for passersby. When they meet pedestrians, they would toast a bowl of rice wine and warmly invite the guests to a banquet at home. Those invited to the banquet will feel very proud and honored. When the guests are satiated and ready to leave, the host will present the guests food and wine. On the second day of the lunar new year, all households bring delicious food and their cattle to the field. They feed the cattle with wine, burn incense, kowtow, and symbolically let the cattle plow the field, in hopes of a good harvest in the coming year. After the ceremony, Pumi people gather for a potluck lunch.

Chapter Two   The Inheritance of Wine Making Technology

The history of brewing and drinking of ethnic minorities in Yunnan can be traced back to the Warring States Period, and the long-standing wine culture is also reflected in their unique brewing methods, the rich varieties and the unique flavors of wine. Fruit wine, water wine, distilled wine (liquor), milk wine, mixed wine, etc., rich in not only ethnic characteristics, but also regional characteristics, all embody the long-term, creative labor experience and wisdom of Yunnan ethnic minorities.

Section 1   A Brief History of Wine Making among Ethnic Minorities

Discovery and Utilization of Distiller's Yeast

With the development of primitive agriculture, mankind began to have surplus grains. However, due to the limited storage conditions, the remaining food had repeatedly deteriorated. Some went rancid and spoiled, while some became edible naturally fermented wine. With the increase in productivity, the ancestors of ethnic minorities began to observe the process of food turning into wine, explored methods to promote grain fermentation, and used various liquor medicine1 to produce natural distiller's yeast. According to the creation epics and myths of various ethnic groups, those who discovered Jiuqu are mostly hunters and male ancestors engaged in farming. Therefore, it can be inferred that ever since the late matriarchal clan period or after the formation of the patrilineal clan society, ethnic minorities had started collecting and using herbs that can promote the fermentation of gains

---

① Liquor medicine: The general term for the herbal raw materials used by Yunnan ethnic minorities to synthesize distiller's yeast.

into wine.

The Yi language classic "The Beginning of All Things", inherited and spread in the Ailao mountainous area of Yunnan, records that the ancestor of Yi people, Se Separ, discovered the principle of wine-making from the rancid rice. Although he spent all his life searching for the right herbs, he failed to make distiller's yeast. His apprentice Huolo Nijiu took over the hard work, relying on collective wisdom and strength, and finally found the herbal raw materials and the methods for synthesizing distiller's yeast. The epic "The Beginning of All Things" vigorously depicted the untold hardships experienced in searching for the wine herbs:

The wine is made by everyone.

Se separ is the ancestor of wine-making.

Boil buckwheat grains with clean water from ninety-nine springs.

The water is full of dew from ninety-nine kinds of flowers.

The utensils used to make wine are hollowed out of cedar trees.

The distiller's yeast is made of sixteen kinds of herbs

Which were found by hundreds of feet.

Huolo Nijiu is the ancestor of the wine industry.

He led the crowd to climb mountains and clouds,

Trod out ninety-nine paths in the world.

The buckwheat are the hard work of everyone,

The herbs are the sweat of everyone.

The distiller's yeast is the achievement of everyone.

Many ancestors of wine-making did not leave a name,

Wine-making is the wisdom of everyone.

Distiller's yeast was not discovered by a single person overnight, but explored by people in the long production practice. With beautiful and elegant language, this epic created poetic, picturesque episodes of ups and downs which demonstrated the pure and sincere ideology of Yi ancestors that "work creates beauty and labor creates life". Although the epic pointed out that "16 kinds of herbs are used to make distiller's yeast," it did not specify the herbs. According to the ancient book "Roots and Origins: Song of Wine Herbs", found in Luquan and Wuding, Yunnan, distiller's yeast was made of twelve kinds of herbs. The book listed the herbs and explained the synthesizing methods:

Twelve kinds of herbs were used in ancient times.

Six kinds are on rocks.

Dig them there.

Six kinds in mountains.

Dig them there.

Dug back from rocks,

Dug back from mountains,

Twelve kinds of herbs put together.

Pounded and sifted,

Mixed with barley flour and water,

Pinched into small balls.

Covered for seven days and nights,

Uncovered and dried,

There comes distiller's yeast.

First, "tousled hair",[1]

Second, scutellaria baicalensis,[2]

Third, gentian,[3]

Fourth, Bupleurum,[4]

---

[1] "Tousled hair": Nostoc flagelliforme, herbaceous plant with fragrance, shaped like hair; roots are used.

[2] Scutellaria baicalensis: roots used, sweet taste.

[3] Gentian: roots used, bitter taste.

[4] Bupleurum: leaves used, spicy and sweet taste.

Fifth, Rubia,[1]

Sixth, a handful of fragrance,[2]

Seventh, Lan Gou,[3]

Eighth, ground melon,[4]

Ninth, Amomum tsao-ko,

Tenth, Tigou,[5]

Eleventh, chili powder,

Twelfth, Aconitum kusnezoffii Reichb.[6]

Twelve kinds of herbs in total,

Cooked into good distiller's yeast.

Since the Song Dynasties, the brewing industry of Yunnan ethnic minorities has made great progress, and the brewing and drinking of water wine have become very common among folks. At the beginning of the Yuan Dynasty, the Italian traveler Marco Polo traveled to Yunnan. In his "Marco Polo's Travels", he mentioned the situation of Yunnan's brewing industry many times. This shows that during the Song dynasty, all ethnic groups in central Yunnan have been using distiller's yeast at a very proficient level. In the Ming Dynasty, there appeared people who were specialized in synthesizing distiller's yeast. The famous traveler Xu Xiake roamed among the mountains and rivers of Yunnan, rambling along the ancient tea-horse road, from Nanhua County of Chuxiong Yi Autonomous Prefecture of Yunnan into Xiangyun County of Dali Bai Autonomous Prefecture. "Down the mountain, past a village, heading north for 1000 meters, up a hill, 1000 more meters, past a lake, there are several households on the northern hill, which is called the village of liquor medicine."

---

[1] Rubia: red roots used, sweet taste, with the functions of relaxing tendons and activating blood circulation.

[2] A handful of fragrance: wild peppermint with fragrance, flowers used.

[3] LanGou: herbaceous plant which helps digestion.

[4] Ground melon: also known as Ficus tikoua Bur, fruits used, sweet taste.

[5] Tigou: leaves used, sweet and fragrant.

[6] Aconitum kusnezoffii Reichb: root used, bitter taste.

The address of such village is unknown now, but considering the name of the village, it can be speculated that the production of distiller's yeast in Yunnan had reached a certain scale during the Ming Dynasty.

Since the Ming and the Qing Dynasties, with the development of pharmacy, ethnic minorities have accumulated profound knowledge about the herbs used to prepare distiller's yeast. Many ethnic groups have been able to make wine of varied flavors, colors and lustre to cater people's diverse tastes by adjusting the proportions of certain herbal components in distiller's yeast according to the raw materials used for brewing. Take the Lisu people in the Nujiang Gorge of Yunnan for instance. They use gentian as the main raw material to prepare distiller's yeast. The method is to crush the gentian into a dough, steam it, and cover it in a bamboo basket for several days. When the gentian dough is fermented, the distiller's yeast of Lisu people is ready. The Lisu folk song "Qinggong Tune" passed down from generation to generation sings:

Don't be sad about steaming wine,

Don't be worried about steaming wine,

I pick up my basket,

Pick up my basket.

Take a look on the mountain ridge,

Take a look on the hillside.

I walk over the mountain ridge,

Down to the hillside,

I see a meadow of herbs,

And the sward of gentian.

I bring the herb back and give it to you,

Bring the gentian back and pass it to you.

You pound the gentian into pieces,

And knead the herb into dough,

Steam the gentian for three days,

Ferment the herb for seven nights,

Expose it in the sunlight,

Dry it in the moonlight.

After three days of exposure,

After three days of drying,

My dear girl can steam wine,

My dear girl can make wine.

Same as Lisu people, the Nu people in Biluo Snow Mountain and Gaoligong Mountain are found of liquor and have mastered the method of making high-quality distiller's yeast early. The main ingredients used by Nu people are corn and gentian. First, mash the gentian, soak it in cold water for a whole day and night. Mix the red and bitter liquid with cornmeal. Knead the dough into egg-sized balls. Then, layer the balls in a basket, and sprinkle rice bran between the layers to prevent adhesion. Finally, put the basket near the fire pit for fermentation. Dry the fermented balls near fire, and the distiller's yeast is ready to use.

The yeast for the Menguo liquor of Hani people is prepared with more than 20 kinds of ingredients such as tree roots, bark, leaves, fruits, spices, rice and flour. All of the ingredients must be finely ground for their respective unique fragrances. The finer the grinding, the richer the aroma, and the better the yeast. Mix the finely ground ingredients with the fully fermented dough, knead the dough into pan cakes and dry them for future use. However, Hani people do not use up all the distiller's yeast at one time. They always put away a little bit which is called "mother of yeast" and would be used for making new yeast in the future.

Jingpo people would hold an herb-picking ceremony before picking herbs for distiller's yeast. Every year, during the ninth and tenth lunar months, the village would select a pair of good looking and virtuous youth for the ceremony. Led by the priest and prestigious elders, they bring rice wine, eggs, and glutinous rice to the mountain and sacrifice the food to ancestors in the clearing. Only after the priest has sung the ritual song inherited from ancestors, can he lead people to gather herbs for brewing wine. Jingpo people believe that the grander the herb-picking ceremony is, the better the quality of the yeast.

The raw materials used for making distiller's yeast vary greatly due to the geographical and cultural differences of ethnic minorities. Lahu people stir fry Chinese Thorowax Root,

Cinnamomum hupehanum bark, banana peel, orange peel, grass roots, stalks and fruits of certain spicy plants together in an iron pan, then pound them before mixing them with the distiller's yeast saved for making new yeast. Seal the mixture and cover it with straw for fermentation. Then fermented mixture would be used as distiller's yeast. The distiller's yeast of Tibetans is unique. It is made of a plant called "Muduzige". Tibetans would mix the bile of fish, goats, bison and other animals with the plant, grind the plant. Then add some flour and cold water, knead the mixture into pan cakes. Dry the cakes before using them for brewing barley wine. Yi people use the most ingredients to prepare distiller's yeast. For instance, the Yi people in Luquan and Wuding often use the following ingredients: cinnamon leaf, Cochinchinese Asparagus Root, wasp, pine root, turnip, scutellaria, gentian, mint, Radix Aconiti Kusnezoffii, pepper, malt, Nostoc flagelliforme, Sambucus Japonica Reinw, Codonopsis Radix, Tuber Fleece-flower Root, moso bamboo, Gastrodia elata Blume, Amomum tsao-ko, honey, wheat flour, buckwheat flour, cornmeal, etc. However, brewers can make appropriate adjustments to the yeast ingredients, proportions and preparation procedures according to the raw material used for brewing, seasons, their taste and color preferences, etc. The distiller's yeast used by Hani people for brewing Menguo liquor also varies in ingredients due to the differences in regional environment. For example, the Hani people in the northwest like to add more tangerine peel, cinnamon, and fennel seeds, while those living in the southwest prefer to add more pepper, fennel, and Artemisia apiacea etc.

Development of wine making technology

During the Warring States period, Banxunman's wine-making technology had reached a very high level. "The Book of the Later Han Dynasty. Biography of Southern Ethnic Groups . Banxun Tribe" records that Banxun tribe once swore with Qin with sake: "If Qin offended Banxun tribe, Qin would lose a pair of dragon made of Huanglong jade; if Banxun tribe offended Qin, Banxun would lose a jar of sake." The "sake" invented and brewed by Banxun people was the best liquor at that time. This ancient and traditional wine-making technology is still the main brewing method of the famous Japanese "sake". The descendants of Banxun gradually merged into the Han, Miao, Yao, and Tujia nationalities and became the mainstream of the modern Ban clan, mostly scattered in Yunnan, Sichuan, and western Hunan today. It could be inferred that during the Warring States Period, the ethnic minorities in Yunnan had mastered wine making technology.

During the Sui and the Tang Dynasties, the Dali area of Yunnan had developed relatively advanced brewing technology. The Heman people in the Erhai Lake area were extravagant when they got married. Dozens of jars of wine would be used at weddings, which implies that the output of wine at that time was not low. They made wine for entertaining guests on November 1st every year... For three days, they rejoiced at the celebration pursuing nothing but pleasure. During the Nanzhao period, the customs of entertaining guests, drinking wine and having fun during festivals had already existed among the various ethnic groups in Yunnan. "Tai Ping Yu Lan"(Song Dynsaty) records that the Moxie people had the custom of drinking and dancing. But their skills of using distiller's yeast were not good enough. As Fan Chuo said, the wine made of rice tasted sour.

In the Song Dynasty during the Dali Kingdom period, Yang Zuo, a native of Sichuan province, came to Yunnan to buy horses. In a place 150 miles away from Yangju City, the local king Shumi gave him a warm reception with "Tengzi wine". In fact, this kind of wine is the "Uncaria wine" popular in the ethnic minority areas of Yunnan. In the third year of Shaoxing in the Southern Song Dynasty (1133), minorities from the southwest went to Luzhou, Sichuan to sell horses. Nearly 2,000 people went there. Among the cargo shipped there was wine, which proves that the wine produced by ethnic minorities in Yunnan has been transported to the inner regions ever since the Song Dynasty.

In the Yuan Dynasty, when brewing wine, ethnic minorities of Yunnan often add spices to improve the taste. When Marco Polo traveled to Yunnan, he saw Kunming people use grains to make wine. When spices were added, the wine was fragrant and delicious. In Yongchang, western Yunnan, he also saw rice wine mixed with many kinds of spices, and praised it as an excellent type of wine.

"The Folklore and Customs of Southwest Minorities" of the Ming Dynasty recorded that in Southern Yunnan, tea served is grain tea while wine served is liquor. "Jiajing Dali Chronicles" contains a poem by Chen Benli drinking in Dali: "Golden cups for Haraji, silver tubes for Zaluma; drunk in the restaurants by the river every day, forgetting that I am far away from home". The "Halaji" mentioned here is liquor. In "Principles of Drinking and Dining" by Hu Sihui, it was called Alaji liquor, which, according to research, is the transliteration of the word "Arradk" in Southeast Asia. It is an alcoholic beverage distilled from fermented coconuts. This shows that during the Jiajing period, Yunnan had developed relatively advanced distillation technology. "Zaluma" is the Uncaria wine popular in ethnic minority areas. "Jiajing Dali Chronicles" records the method of making Uncaria wine: "The wine is made in a crock. When it is ready, put Uncaria in the crock, add boiled water and simmer the wine. Suck the wine with the Uncaria. It tastes better than ordinary wine, and is call Zaluma." Therefore, Uncaria wine is also known as "Zaluma" wine, which is characterized by "holding the Uncaria to drink it". Uncaria is a kind of shrub that is hollow, so it could be used to suck the wine. "Sketch of Dian: Production" has a more detailed record: " Uncaria, which is vine, can be used to make wine. The aboriginals brew rice wine in urns. When it is ready, put vines in the urn, add some water and stew. Hosts and guests will suck the wine with the vines. Uncaria can cure cholera and relieve many kinds of toxins. Its effect is the same as that of brown coconut." Guests and hosts gather around the wine crock and take turns to suck the wine. Uncaria is actually a kind of condensing tube. The wine begins to volatilize when heated, and condenses through the Uncaria tube. This process indicates that such wine should be distilled wine, namely, liquor. Its history can be traced back to the Dali Kingdom period. Nowadays, drinking with Uncaria tubes is still popular among the Miao and Zhuang nationalities in southern Yunnan.

Thanks to the mature distilling technology, some kinds of famous liquor appeared in various parts of Yunnan in the Qing Dynasty. "Yunnan Chronicles" records that during the Yongzheng period, there were several kinds of liquor, white liquor and yellow liquor in Yunnan. "Dianhai Yuheng Chronicles" records various kinds of famous liquor in the Qianlong period, for example, the Lishi liquor (Lishi: heavy rocks). Lishi liquor, produced in Dingyuan, is also made from sorghum. The person who named the liquor said that the effect of such liquor was as heavy as rocks. Sorghum liquor produced in Yuanmou in central Yunnan has a taste of the Ganshao (steamed wine) produced in the north and is obviously a kind of liquor. During the same period, Wuding's Huatong liquor, Kunming's Nantian liquor and clove liquor from various places were very famous, and Dali's Heqing liquor had a more mellow taste than Fen liquor.

After the Qing Dynasty, new brewing techniques were quickly spread among ethnic minorities in Yunnan. Shaoxing liquor imported from the mainland was brewed with water from the well of Wu in Kunming. What came from Burma was Guci wine. Kaihua (today's wenshan) produced exotic wine and stored in glass bottles, which should be an imitation of foreign wine. Undistilled wine is called "white wine". For instance, "The Sequel of Dian Trips" says that glutinous rice is made into sweet wine, commonly known as white wine, which is still popular among folks nowadays.

Section 2 Wine Making Legends

The Han nationality has legends about wine gods in charge of wine, different ethnic minorities also use various legends to interpret the origin of wine. These legends can be divided into two categories. One is the theory of accidental discovery. According to Yi legends, people did not know the existence of the magical beverage of wine until they discovered that the wild grapes used to worship the mountain gods can be intoxicating after fermentation. The water wine of Wa people was also an accidental discovery. Wa women would wrap some rice in banana leaves for lunch when they worked outside. As the rice was accidentally left on the branches of trees near the field, it was naturally fermented into rice wine. The other is the theory of gifts from gods, which is full of magical and charming imagination. Nu people believe that wine is a wonderful drink bestowed by gods. The gods gave three foods to Nu people: "Cuoque" (Vinegar wine), "Cuola" (liquor), and "Cuoren" (corns), which collectively referred to as "Cuodong", two of which are wine. Sulima of Pumi ethnics was secretly learned by the ancestor Shizhuan Heda from the monsters at the risk of his life. The creation epic of Yao ethnics, "Mi Luotuo", says that the ancestor of mankind, Mi Luotuo, is a demigoddess. After she has created everything in the world, she began to create humans. She used rice to create humans, but it turned into wine. In the creation myth Lahu people "Dupa Mipa", the earliest wine was in the charge of god Esha, and the appearance of everything on earth is inextricably linked with wine:

.....

The cold season came,

The hot season came,

The fruits were also ripe,

But without sweet taste.

In Ersha's wine,

There seemed to be five flavors,

Sour, sweet, bitter, spicy, and fragrant.

Esha told clouds to sprinkle into the sky the wine,

Which turned into rain and fell from the sky.

The rain fell on fruits,

The fruits became sweet and fragrant,

The fruits keep growing day by day,

Which delighted Ersha.

.....

The creation epic "Lebao Zhaiwa" handed down from generation to generation by Jingpo people, chanted the origin of wine in a special way:

My son Ning Guandu,

The ancestor of tribe chiefs!

The milk bestowed by your father Jinen,

Has been squeezed between the sky and the earth;

The milk granted by your mother Weichun,

Has been sprinkled on earth.

Where Jinen milk and Weichun milk will be sprinkled,

A variety of plants for distiller's yeast will grow.

The wine brewed with the yeast will be mellow and fragrant.

As for your father's Jienen milk,

You all can share;

As for you mother's Weichun milk,

You all will taste.

Ning Guandu mentioned in the epic is the founder of the Jingpo nationality. His father is named Penggan Jinen, and his mother, MuzhanWeichun. When Ning Guandu was crossing the wilderness before leaving his parents, he listened to his parents' admonitions. The first admonition is about how to find a residence; the second is to venerate the heaven; the third is to venerate the earth; and the fourth is about how to make wine. His parents reiterated that fragrant wine was transformed from the breast milk of the parents.

Derived from the same origin of the wine culture of Han, the legends of ethnic minorities have it that wine is bestowed by gods in heaven. Ancestors' risking their lives to elicit wine-making techniques from monsters reflects the difficult process of fighting against the sinister natural environment (Pumi myth). The story of goddess Mi Luotuo, who intended to make people with rice, but created wine instead, is a review of the history of the people whose survival and development were built on the rice culture (Yao myth). Our ancestors have known that wine-making plants formed from parental milk grow in the mountains (Jingpo myth), and that all creatures in the world became spiritually lively, charmingly aromatic and attractively sweet when gods sprinkled wine from the sky (Lahu myth). Through the halo of primitive culture, we can feel the sincere joy of the ancestors who discovered wine at the early stages of human civilization.

These myths, legends and epics about the origin of wine are an essential part of the ethnic wine culture. Reading carefully, we can see the important role of women in wine culture. It is females who first discovered wine and engaged in wine-making activities. First, among the many myths and legends of various ethnic groups, there are two kinds of people who discovered wine, one is women, and the other is hunters. But the legends about discovery done by women account for the vast majority. The wine of Nu people is bestowed by immortals are stalactites of female images. The wine of Jingpo people is made from the milk of the mother of the ancestor Ning Guandu. In addition, many ethnic groups, such as Yi, Lisu, Pumi, and Wa, have legends of wine formed from the deterioration of leftover rice. In this type of legends, almost all the people who found the leftover rice turned into wine are females.

Second, in the creation epics and legends of ethnic minorities, the first people engaged in wine-making are all females. Miluotuo, the ancestor of the Yao nationality, is the female who used rice to make people, but turned it into wine. Wa people recorded the origin of wine in the creation myth "Si Gang Li" which affirms that Wa people learned the technique of brewing from bees, but the first person to be able to brew is Ya Dong, the female leader of the Wa society. "Hani Apei Congpopo", the migration epic of the Hani people living on both sides of the Red River in Yunnan, sings the development of Hani in a comprehensive and systematic way. After describing the origin of human beings and their learning to use fire, the epic recalls the long journey from primitive hunting to domesticating livestock and farming civilization that the ancestors have gone through under the leadership of a female named Zhenu. With the development of farming civilization, Hani people produced inexhaustible rice and started to make wine with grains:

(Excerpt from Hani Apei Congpopo)

Hani has a talented person, Zhenu

Who is well-known everywhere.

She picked plump grass seeds,

Planted them into the loosest and darkest soil;

She fetched water from the lake,

Splashed water on the seeds just like the rain god.

The grass seeds sprouted stout buds,

The grass seeds grew tall stems,

Which were full of golden seeds.

The ancestors ate the fragrant seeds,

And named them corn, millet and sorghum.

Zhenu's harvest went up and down.

Realizing her planting was not consistent with the right season,

She invited Zhesi, the shepherdess of cattle and sheep.

The girl Zhesi had an idea:

She pointed to twelve animals,

And set the year and month zodiac signs,

One year was divided into twelve months,

Thirty days and nights per month.

Hani's calender starts with mouse,

And ends with corpulent pig.

Planting in accordance with the zodiac,

Zhenu grew inexhaustible grains,

With which she brewed fragrant wine,

That became Hani's inseparable partner.

Third, in modern and contemporary times, it is ethnic females who are engaged in wine-making. For instance, in Jingpo villages, wine making is the most basic life skill for women. Jingpo women start learning to make water wine and Shaojiu from an early age because this is an important criterion for evaluating their viability and labor skills. The day after the wedding, the most important thing for the Jingpo bride is to make fine wine and serve her parents-in-law. If she fails to make good wine, she will be laughed at for life. In the Pumi wine song "Sulima", verses like "mother and sisters are busy making wine, grandpa and father are enjoying wine" are repeated many times. When looking for the right person, Lisu young men expect their future wives would be kind-hearted, hardworking and capable, so that the new family has "someone who cooks in the morning and serves wine in the evening". At Wa weddings, the elder people bless the newlyweds "May you have a daughter to distill liquor and cook meals. May you have a son to plow the land." Such blessings reflect the division of labor and the inevitable wine-making responsibility of women.

Section 3　Types of Wine and Brewing Methods

The first type of wine that the ancestors of ethnic minorities have discovered and enjoyed is fruit wine, whose fragrance suffused for thousands of years. With the continuous development of agricultural production, wine brewed with grains has entered people's lives. Among them, water wine has become a splendid movement in ethnic minority wine culture due to its long history and far-reaching influence. In the long course of brewing and drinking, people gradually understood and made good use of the effect and function of liquor to create a variety of compound liquor, which play a unique role in enriching life and improving people's health.

Fruit wine and its brewing method

Shutou Wine

(wine made on branches) As early as the Yuan and the Ming dynasties, in the tropical and subtropical forests of XishuangBanna and Dehong, ethnic minorities worshiped water and were found of wine. There were trees, seemed like palm trees, with 8 or 9 the stems on the top. People cut the tip of the stem and tied a gourd ladle to it. After a night, there would be a ladle of wine, fragrant and sweet. People would be intoxicated when drinking it. The wine would decay overnight; if distilled, one can only drink up one cup (Ming. Qian Guxun "Biographies of Ethnic Minorities"). During the Tianqi period of the Ming Dynasty, based on textual research, Liu Wenzheng in his "Dian Chronicles·Volume Thirty-two·Additional Production" depicted Shutou wine: "aboriginals in XishuangBanna and other places put distiller's yeast in crocks and hung the crocks under the fruits. There would be wine as the fruit juice was extracted and poured into the crocks. The tree leaves are like Pattra leaves and are used to write Buddihist scriptures". At the beginning of the Qing Dynasty, the method of directly extracting the juice from the fruits of palm trees to make Shutou wine is still commonly seen in the authoritative official documents. For instance, "Yunnan Chronicles·Tusi" of the Kangxi period has the following description: "The aboriginals put distiller's yeast in crocks and suspend the crocks under the fruits. Cut the fruits for juice to drip into the crocks so as to make wine, named as Shutou wine." According to the textual research of "New Compilation of Yunnan Chronicles· Products Research", the tree used to make Shutou wine is tropical coconut palm. Because of the sugar content, its juice can be fermented into wine. Ethnic minorities did not pick the coconuts, but put distiller's yeast into gourd ladles, crocks and tea pots, hung them under the fruits, and cut or drilled small holes on the fruits for juice to drip into the containers to ferment into wine. When traveling in Yunnan, Tan Cui was amazed by this natural wine-making method, and listed it in "Wine Chronicle" and "Botany" in his "Dianhai Yuheng Chronicles". At the end of the Qing Dynasty and the beginning of the Republic of China, the method of making wine on branches still remained among the ethnic minorities in western and southern Yunnan. However, it is rare nowadays.

Wine made of grapes

"Dianhai Yuheng Chronicles" records that "The grapes produced in Yunnan are good, but locals don't know how to make wine. All grape wine are from Zhongdian which is connected to Tibet and has many Tibetans residents". This shows that as early as the mid Qing Dynasty, the Tibetans in Zhongdian, northwestern Yunnan, have already produced grape wine. After the late Qing Dynasty, French missionaries entered Deqin in northwestern Yunnan, bringing in the French grape variety--- rosy honey and wine-making techniques to the snowy plateau, enriching the local wine craftsmanship. Today, in the Cizhong village of Deqin, the vines, planted by the missionaries hundreds of years ago, are still graceful with luxuriant foliage and abundant fruits. This kind of wine made of rosy honey grapes with French wine-making techniques has gradually developed into a uniquely charming brand "Yunnan Red".

In addition, as early as the Qing Dynasty, there have been a variety of fruit wine in Yunnan, such as mulberry wine, Roxburgh wine, hawthorn wine and snow pear wine. Now these kinds of fruit wine have become a must-have for festival celebrations and visiting relatives and friends.

Water wine and its brewing method

Water wine is fermented wine made of raw materials such as millet, wheat, rice, etc. mixed with distiller's yeast. Drinking both the juice and the dregs of the grains together, the ancients called it "wine mash". With the most varieties, water wine has become the most popular type of wine among Yunnan ethnic minorities. The "sweet wine" of Zhuang and Achang people, the "purple rice wine" of Hani, the "grain wine" of Yao, the "barley wine" of Tibetans, the "Yin wine" of Naxi, and the "Sulima" of Pumi people, all of them fall into this category. In many ethnic minority areas, fermented wine is also called white wine, and according to the degree of fermentation, it is divided into two types: sweet white wine and spicy white wine.

Sweet white wine is made of grains such as rice, corn, millet, and so on. Boil or steam the grains. When the cooked grains cool down, mix them with distiller's yeast and a little amount of cold water. Store the mixture in crocks in a warm and dry place. Sweet white wine would be ready in 1-2 days in summer and 3-5 days in winter. Lahu people use glutinous rice as raw material. The brewing method is as follows. Boil glutinous rice and rice bran for a while before steaming them with a wooden steamer. Put the steamed glutinous rice and bran in a clay pot, sprinkle some homemade distiller's yeast. Store for 1-2 days, the sweet white wine will be ready to drink, cool and delicious. Sweet white wine is essentially the water wine formed when the starch in the grains is completely saccharified, and the alcoholization process is about to begin. It is sweet and delicious, emitting a bit of faint alcoholic aroma, and is a suitable beverage for young and old.

Various ethnic groups have a long history of brewing sweet white wine. As early as the Yuan and the Ming dynasties, it has been commercialized. At the beginning of the Ming Dynasty, while traveling from Dali to Yongchang (now Baoshan), Xu Xiake passed through a mountain valley, called Wanzi Bridge, with a few families on the southern side. There were pulp sellers. People drank the pulp together with the dregs. It is the water wine brewed locally. It can be seen that as early as the Ming Dynasty, even in the mountain valleys, sweet white wine has become a commodity for the rushing business travelers on mountain paths to "sip it with the dregs". Sweet white wine has high nutritional value. Boiled eggs with sweet white wine is a good treat for guests of ethnic minorities. It is also health-care food used by puerperal women to nourish the body, restore vitality, and promote lactation. Serving guests with such delicious food has become a custom since the Ming and the Qing Dynasties. Tan Cui, the governor of Luquan during the Qianlong period, was deeply impressed by the local people in this area. They made white wine each year during the middle ten days of the twelfth month of the lunar calender. When guests came in the new year, they would present a bowl of boiled eggs with sweet white wine to demonstrate their intimate relationships. To this day, during the festive season, soaking rice and steaming rice to make white wine is still one of the most important pre-holiday preparations for many ethnic minorities.

Spicy white wine is low alcohol content wine made from rice, glutinous rice, corn, wheat, barley, millet and other grains. The ethnic groups in Yunnan have a long history of brewing water wine. In the literature and classics of the middle and late Ming Dynasty, there are records about making distiller's yeast and brewing spicy white wine.

Spicy white wine of Yi

Yi people are good at brewing spicy white wine. Glutinous rice is the preferred raw material, and grains such as rice, corn, sorghum, millet, etc. can also be used for brewing.

The basic steps of making spicy white wine are as follows:

1.Soak or cook raw materials.

2. Put the soaked or cooked grains in wooden or bamboo Zengzi (Chinese traditional steamer) and steam it thoroughly. The steamed grains at this time are called "wine rice".

3. Spread the steamed wine rice on a clean bamboo mat so that the rice will cool down naturally.

4. Sprinkle distiller's yeast and some cold water on the cool wine rice, stir to mix them evenly. Then store the mixture in clean crocks. For clean and pure juice, put a bamboo filter at the bottom the crock in advance, so that the juice and the dregs could be separated.

5. Once the wine rice is stored in the crocks, the saccharification of the starch in the grains will be completed within 1-2 days to form sweet white wine. 5-7 days later, owing to the function of the yeast, the fermentation will be completed to form spicy white wine with rich aroma. Now, it can be taken out for drinking or storage. Since the temperature of the wine rice must be kept at a certain degree after being crocked to facilitate fermentation, the crocks are often placed close to the fire pit, or buried in rice bran. In winter, even quilts are used to cover the crocks. Therefore, this process of brewing is also called "muffling white wine." When the white wine is ready, take the crocks out of the rice bran or the quilts, which is called "getting out of the nest".

6. Enjoy the spicy white wine after getting it out of the nest, either drink the original juice, or add in an appropriate amount of cold water according to the alcohol content or your favor of taste. If you don't drink it temporarily, seal and store it immediately, either store the original juice and dregs, or filter the juice for storage, or mix it with an appropriate amount of cold water for storage. Stored in summer, it can be kept for about 20 days. Stored in winter, it can be reserved for several years. The longer the storage time, the more mellow the spicy white wine tastes. The so-called "Yi family old wine" is this kind of water wine stored for a long time.

The water wine stored over years has a lasting aroma. Its dregs and juice have been completely separated. The juice is clear and bright, slightly yellowish brown. Though mellow and refreshing, even drinkers can't resist its strength, and would be intoxicated with 2-3 cups. Presenting a bowl of old spicy white wine is one of the highest etiquette manners for Yi people to receive elders and good friends.

Yinjiu of Naxi

The ancestors of Naxi people loved wine. The Dongba scripture "Er Zi Ming (a calendar of diet)" is an ode to agricultural production and labor. The first part of the long poem describes the whole process of growing wheat and making wine. It shows that Naxi people were able to make wine in ancient times. Yinjiu is the local water wine originally created by Naxi people. It dates back to the Daoguang period of the Qing Dynasty. Yinjiu is made with high-quality rice produced in western Yunnan, and is brewed according to the traditional fermentation method. After soaking, steaming, and cooling the rice, add distiller's yeast to saccharify the wine rice. Keep the wine rice in crocks for low-temperature fermentation for about one month before pressing the juice out. Then seal and store the juice for aging. Yinjiu is amber, limpid, sweet and mellow, rich in glucose, a variety of amino acids and vitamins, and has a good nourishing effect.

Purple Rice Wine of Hani

The purple rice wine brewed by the Hani people in the valleys of southern Yunnan and on both sides of the Red River is fermented with high-quality purple rice produced locally. It is the best drink for receiving guests and has been famous as early as the late Qing Dynasty and the early Republic of China. In addition, Dai, Yi, and Jingpo people also have a tradition of brewing purple rice wine. The brewing method is the same as that of Yi people's water wine.

Lajiu of Lisu

The water wine "Lajiu" brewed and drunk by the Lisu people in the Nujiang Gorge of Yunnan is named after it unique drinking method. Lisu people use wheat, corn, sorghum, etc. as raw materials. Cooked and steamed, they are mixed with distiller's yeast and stored in crocks for fermentation. When guests arrive, Lisu people will take out a proper amount of fermented wine dregs and stew it over the fire pit. Guests sit around the fire pit in a circle. In the mean time, the host keeps adding water to the wine pot, pulling(La) and filtering the dregs out while pouring Lajiu for guests until its taste fades away.

Bamboo Wine of Dulong

Whenever the harvest season comes, every Dulong household makes wine. Instead of using crocks, they brew with bamboo tubes. First, choose the best bamboo in the mountains to make wine tubes. Then mix the cooked rice, wheat or sorghum with distiller's yeast and store the mixture in the bamboo tubes. After seven days, it will become fragrant and ready to drink. Such wine is sweet and delicious, and has a slight aroma of bamboo.

"Bulailon" of Wa

Wa people call water wine "Blailong", which means "new year wine". Water wine is a traditional refreshing drink used by Wa people to dissipate heat, detoxify, drive fatigue, and strengthen the body. Its raw materials are mainly red rice, corn, sorghum, wheat, buckwheat, glutinous rice and other grains, among which red rice and sorghum are the best choices. The brewing method of Bulailon is as follows. Stir-fry the raw materials and grind them, sprinkle a small amount of water and mix well before steaming it. Let the steamed grains cool down, add a proper amount of distiller's yeast and mix evenly. Put the mixture into the baskets covered with banana leaves, then leave the baskets near the fire pit for the mixture to ferment. Afterwards, store the fermented grains in crocks and seal them for several months. Wa people have known for a long time that the longer the wine is reserved, the sweeter it will be. They would not drink the Bulailon if it has been stored for less than one year. Before drinking, add some spring water and soak for about 10 hours. This is why Wa people call water wine "soaked wine". Soaked and filtered, Bulailon be ready.

Water Wine of Dulong

The Dulong people in the Dulong River Basin in northwestern Yunnan are fond of water wine, and their brewing methods are quite unique. They mostly use corn, rice and sorghum to make water wine. Cook or steam the ground raw grains thoroughly, then let them cool down. Add distiller's yeast to the steamed grains and mix well. Dig a crock-shaped cellar on the ground. Pad the bottom and walls of the cellar with clean banana leaves, then place the mixture in the cellar. Cover the wine rice with banana leaves to completely isolate it from the soil layer, then seal the cellar opening with mud. Burn a fire on the cellar opening to make the wine rice in the wine cellar ferment for 3-4 days. Then, take out the fermented wine rice, store and seal it in crocks for a few days. Before drinking, add water, stir and heat it; or mix it with cool mountain spring water so that the sweet and mellow water wine will relieve summer heat and thirst.

Sulima of Pumi

Sulima wine is the traditional fine wine of the Pumi people in northwestern Yunnan. It is not only a necessity for activities such as worshiping gods and ancestors and exorcising evil spirits, but also an important gift for guests and friends. Sulima is brewed with high-quality barley. When brewing, cook the barley first, then let it cool down before sprinkling distiller's yeast. Stir and mix evenly, then place the wine rice in bamboo baskets for fermentation about 3 days. Store and seal the fermented wine rice in crocks for at least about half a month. The longer the storage time, the better the color, fragrance and taste of Sulima. Generally, when opened 3-4 months later, the crocks are already full of wine with aroma overflowing. Pumi people will add some sweet spring water, soak it for 1-2 days before drinking. The Sulima soaked with water for the first time is called "head wine", which is offered to distinguished guests from afar. The Sulima soaked for the second time is called "second wine", which is served with pork to those who come to help with farm work and house building so as to alleviate fatigue and create a joyous atmosphere. The third soaked Sulima is usually what Pumi people drink for quenching thirst at work. The wine dregs are fed to pigs that are to be slaughtered for the New Year. The delicious bacon and Sulima are the delicacies during the New Year. Pumi people's homemade distiller's yeast contains gentian from the snow-capped mountains and lily from the riverside, therefore, Sulima has a rich aroma and a sweet and cool taste.

Rice Wine of Miao

The Miao people in southeastern Yunnan use rice or glutinous rice to make water wine. Their method is basically the same as that of Yi people. Miao rice wine has high sugar content but low alcohol content. It is the best drink to relieve fatigue and refresh the mind. Miao people often drink it at meals. "Rice soaked in white wine" is a traditional dietary of Miao.

Chestnut Wine of Yao

Yao people are adroit at brewing chestnut wine. In the golden autumn, Yao people harvest the chestnuts and remove their shells. Wash and cook them, drain the water out and let them cool down to about 30 degrees centigrade before adding distiller's yeast. Mix well, and seal them in crocks for two months before serving. Chestnut wine is mellow and tangy, with a mixed flavor of bitterness, sweetness and spiciness.

Barley Wine of Tibetans

Highland barley wine is a traditional Tibetan beverage. It has a very low alcohol content. Its brewing method is very simple. Wash and cook the highland barley first. When the temperature drops slightly, add distiller's yeast, store and seal the mixture in crocks or wooden barrels. After 2-3 days of fermentation, add water and store it for another two days, then it is ready to drink. The barley wine is pale yellow and has a slightly sour taste. The aged barley wine that has been buried for 3-5 years looks like honey, and has a fragrant tangy flavor. Barley wine can quench thirst and refresh, and is a must-have for Tibetan people's weddings, funerals, and festival celebrations.

Emerald Wine of Bulang

The emerald wine is brewed by Bulang people with glutinous rice as raw materials. Its production method is generally the same as that of other ethnic minorities. The difference is that after the glutinous rice is fermented into wine, Bulang people use raspberry leaves to filter out the wine dregs, therefore the wine is limpid and has an emerald color. It is the best beverage used by Bulang people to receive friends and relatives.

Meizi Wine of Manchu

"Mei" is broom corn millet. "History of Wei of the Northern Dynasties", "History of the Sui Dynasty", and " Khitan Chronicles", all have records of Man ancestors making Meizi wine. The method is as follows. Soak the millet in water, then steam it thoroughly. Add water and distiller's yeast to the steamed millet and mix evenly. Seal it in crocks for 2 days before drinking. Some Man people stir-fry the millet, then use the same process to make wine, which is called fried Meizi wine. Meizi wine is sweet and fragrant and is often used in festivals to entertain relatives and friends. Its alcohol content is not high, Man people can booze to their heart's content without harming their health.

Glutinous Rice Wine of Shui

Shui people brew glutinous rice wine by themselves, following their traditional methods in the process of steaming rice, mixing distiller's yeast, fermenting, and stewing. This kind of rice wine does not have high alcohol content, but a mellow taste, so everyone can drink it. Shui people have the custom of brewing wine on birthdays. The birthday wine must be preserved underground for many years, and will be unsealed respectively on important occasions, such as when the person gets married, builds a house, and passes away. Because of its special brewing method and long aging period, it has a long-standing reputation. Having preserved for more than 30 years, such glutinous rice vintage wine will condense into syrup as thick as honey. When drinking, add mountain spring water at a ratio of 1:10, or add a little fresh rice wine to create fabulous flavor.

Gudu Wine of Nu

Being found of wine, Nu people are also good at making wine. The Gudu wine of the Nu in Gongshan mountain is the most distinctive. Gudu wine is brewed with "Gudu rice" (made from cornmeal and buckwheat). The brewing method is the same as the water wine brewing method of other ethnic minorities. Add distiller's yeast when the cooked Gudu rice is cool, mix well before placing it in bamboo basket for fermentation. Three to five days later, store and seal the fermented rice in crocks for more than ten days before it becomes aromatic wine. The special feature of Gudu wine is its drinking method. First, add water and honey. Then, filter out the dregs, and drink its juice. Mellow, sweet, and appetizing, Gudu wine can not only quench thirst, but also nourish the body.

Glutinous Rice Wine with Honey

For thousands of years, Buyi people have inherited and passed down the custom of feasting guests with wine, making friends with wine, celebrated birthdays with wine, expressing their ambitions with wine, and entertaining with wine. Not only do they love drinking, but they are also good at making wine. The rice is grown by themselves, and the distiller's yeast is made from roots of various plants in mountains. Therefore, their homemade rice wine is mellow and sweet. Whenever Buyi people open their wine crocks, the fragrance will fill the room.

Glutinous rice wine with honey is a special beverage of Buyi people. The brewing method is as follows. Steam the soaked glutinous rice. As the steamed rice cools down, add the homemade yeast, mix well, and seal the mixture in crocks. Three days later, add some warm water (15°C), some more distiller's yeast, and a little amount of liquor, mix well, and seal again. Another four days later, add an appropriate amount of honey, mix well and seal again. After 1-2 months, filter out the dregs with gauze, and what is left in the crock is the glutinous rice wine with honey. Fragrant and delicious, this kind of wine has the effects of invigorating the stomach, moisturizing the lungs, eliminating phlegm and relieving cough. As a nutritious drink, it is a good choice for entertaining guests.

Shaojiu and its brewing method

Shaojiu refers to a variety of transparent, distilled wine, generally called liquor. People in different places name it differently, such as Laobaigan, Shaoguo Liquor, Steamed wine, and so on. Shaojiu originated in the Tang Dynasty and gradually became popular after the Song and Yuan Dynasties. Li Shizhen, a pharmacist in the Ming Dynasty, elaborated its brewing method: "Steam the fermented grains. When the vapor rises, collect the drops of liquor with a utensil. Any rancid grain wine can be steamed...It is as clear as water, and its taste is extremely strong, namely the dew of wine." In the Jingtai period of the Ming Dynasty, Zheng Yong recorded in his "Yunnan Chronicles with illustrations VI" that, "Echang ethnics... grow "Shu" to brew wine. They sing and dance when drinking. Dregs are made into cakes which are dried for possible lack of food". The Echang ethnics mentioned here is the Achang minorities in the Baoshan and Tengchong areas of today's western Yunnan. "Shu" is sorghum. "Growing sorghum to brew wine" indicates that this crop is almost exclusively used for wine-making. The account of "making dregs into cakes" shows that the local wine at that time is not the water wine that is "sipped together with dregs", but distilled wine. This fact proves that in the middle and late Ming Dynasty, ethnic minorities in remote mountainous areas have already mastered the techniques of distilled wine.

Small Pot Liquor of Yi

The Yi people in Ailao Mountains are good at brewing Shaojiu. Because distilling is done in small family workshops with small stoves and small pots, it is also called small pot liquor. The main raw materials of small pot liquor include barley, corn, buckwheat. rice, and chestnuts. The process of brewing small pot liquor is divided into two stages. The first stage is to muffle the wine rice. Soak the prepared raw grains thoroughly and cook them. Spread them out to cool. Add distiller's yeast and stir evenly. Store and seal the mixture in crocks for fermentation. The second stage is to distill, that is, to steam the fermented wine rice in a Zengzi (Chinese traditional steamer). The vapor coming out of the wine rice is condensed into liquor by a cooler. The small pot Liquor of Yi is mellow and refreshing. It's a great gift for relatives and friends.

Menguo Liquor of Hani

The Shaojiu brewed by the Hani people on both sides of the Red River in Yunnan is called "Menguo Liquor"(Menguo: stewing pot). It has a long history. The ancient Hani people sang as follows:

What is the custom in August?

The custom in August is Huoxizha (Xingu Festival).

The hot vapor of freshly cooked rice

Pervading among the mushroom houses;

The aroma of Xingu liquor,

Floating in the sky like a cloud.

The raw materials of Menguo liquor are preferably corn, sorghum, rice, and buckwheat, and the brewing process is unique. First, soak the selected raw grains with clean water, steam them and let them cool. Sprinkle distiller's yeast, stir evenly before placing them in a large basket. Cover the basket tightly with straw for the wine rice to ferment. When juice flows out of the fermented wine rice, move the wine rice into a crock, seal the crock with mud made of plant ash. After the wine rice has fermented for another 10-15 days, it is ready for distilling. The wooden Zengzi used for stewing (distilling) wine has a wine vessel placed inside. An iron cooling pot filled with cold water is placed on the top of the Zengzi, and the water in the pot is replaced continuously to keep the pot cool. When the water in the heating pot at the bottom of the Zengzi is boiled, the steam of the wine rice in the Zengzi rises, condenses into droplets at the bottom of the cooling pot, and falls into the wine vessel. Since the distilled liquor is not collected from the outside of the Zengzi as usual, but from the inside of the Zengzi, hence it is called "Menguo Jiu" (stewed liquor). Menguo liquor is crystal clear, mellow and sweet, and is a must-have drink for Hani festivals. In addition to Hani people, Dai, Jingpo, and Lahu people are all good at brewing Menguo liquor of excellent quality .

Steamed liquor of Nu and Lisu

The Nu and Lisu people on both sides of the Nu River in Yunnan call Shaojiu as steamed liquor due to the main process of steaming. The preferred raw material is corn, while sorghum, rice, buckwheat, and millet are also used. The procedures of soaking, steaming, storing and fermenting grains are the same as those of the Small Pot Liquor of Yi people, but the utensils used for distilling the liquor are different. The Zengzi used by Nu and Lisu is hollowed out of logs. A small hole is carved in the middle part of the Zengzi to insert a thin bamboo tube. When pot at the bottom of the Zengzi is heated, the vapor of the wine rice rises and condenses into liquor which falls into the liquor vessel inside the Zengzi, and then flows out through the bamboo tube.

Shaojiu of Bai

Bai people have the custom of "no etiquette is without wine". Brewing Shaojiu is an important sideline of Bai families. After having steamed barley and corn thoroughly, mix them with distiller's yeast made of 48 kinds of Chinese herbal medicine. Store the mixture in crocks for fermentation for 40 to 50 days, then distill it into Shaojiu. The most famous Shaojiu by Bai people is Heqing Qianjiu produced in Heqing County, Dali Prefecture. This liquor has a long history. Its mellow fragrance and delicate sweetness have fascinated people of all ethnic groups.

Stewed glutinous rice wine of Buyi

Stewed glutinous rice wine is unique to Buyi people. The brewing method is as follows. First, make a jar or several crocks of glutinous rice wine. Second, scoop out the liquid and put it in a jar. Put the jar in a hole that is slightly larger than the jar. Place a small bowl at the mouth of the jar, and fill the bowl with water before seal the jar. Then, fill up the space between the jar and the hole with grain chaff or sawdust. Last, ignite the grain chaff to fumigate the jar. Half a month later, the glutinous rice wine in the jar would reduce by two-thirds, and the remaining is stewed glutinous rice wine. Under normal circumstances, 25 kilos of glutinous rice can produce 3 or 4 kilos of stewed wine. Buyi people would store the newly-brewed wine and enjoy it in autumn and winter. The stewed glutinous wine is inky, silky, and mellow. It can quench thirst and invigorate the stomach. It is a good product for nourishing Yin, strengthening Yang and promoting fitness. Due to its complex brewing procedure and high cost, Buyi people would only drink it with their distinguished guests.

Nowadays, many ethnic minorities living in Yunnan, such as Yi, Jinuo, Achang, Hani, Nu, Lisu, Lahu, etc., can brew Shaojiu of different flavors and qualities. The folk song of Pumi people describes very vividly the distilling utensils and the brewing process:

Looking towards the top of the mountain,

There seemed an ocean;

Looking towards the foot of the mountain,

There seemed a rooster jumping.

In the middle of the mountain is a small mouth,

From which the mellow liquor flows out.

Hardworking girl,

Guarded the stove in order to fill the liquor jar;

Greedy young man,

Looked for opportunities to lick the dripping liquor,

While lingering around without intent to leave.

When the liquor jar is full,

The girl was so happy.

She filled the bowl for the young man,

Who was drunk for three days and three nights.

This song describes the arrangement of the stove, the pot, and the Zengzi for distilling Shaojiu as a mountain. The cooling pot at the top of the mountain is compared to ocean, and the raging flame in the fire pit at the bottom of the heating pot is a dancing red rooster, the bamboo tube on the wooden zengzi is a mouth out of which the mellow liquor flows. The short ballad sums up the common characteristics of the Shaochu utensils of ethnic minorities.

Compound liquor

The rich variety of compound/ medicated liquor is an important part of the folk medicine of various ethnic minorities. Taking the advantages of the functions of liquor such as "enhancing the effect of medicine, maintaining beauty, and slowing down aging", ethnic minorities compose medicated liquor to cure diseases and prolong life. In the early Ming Dynasty, the pharmacist Lan Mao (a native of Songming, Yunnan), absorbing the rich medical and cultural experience of various ethnic minorities, compiled a pharmacological monograph "Materia Medica of Southern Yunnan ". In this great book, which is full of unique local and ethnic characteristics, and is more than 140 years earlier than Li Shizhen's "Compendium of Materia Medica", Lan Mao thoroughly discussed the principles and methods of using liquor as medicine, and recorded a large number of secret recipes for composing medicated liquor.

The compound liquor of the ethnic minorities is diversified. Some are formulated with herbal roots, such as Gastrodia elata Blume liquor in western Yunnan, Poria cocos liquor in Ailao mountainous area, Panax notoginseng liquor in southern Yunnan, Cordyceps sinensis liquor in northwest Yunnan, and so on. Some are formulated with fruits, such as Chinese flowering quince liquor, mulberry liquor, plum liquor, Phyllanthus emblica liquor, and so on. Some are formulated with plant stems, such as Gynostemma pentaphyllum liquor, dodder wine and gentian liquor. Some are formulated with animal bones, gallbladders, eggs, etc., such as tiger bone liquor, bear gallbladder liquor, and egg liquor. Still, some are formulated with minerals such as Maifan stone liquor. In terms of efficacy, such compound liquor composed by ethnic minorities are divided into two major categories: one for health-care and the other used as medicine. The health-care type has a wide range of uses and accounts for the majority of compound liquor.

Yanglin Nutritious Liquor

The traditional compound liquor, Yanglin Nutritious Liquor, is well-known at home and abroad. It is named after its place of production, Yanglin Town, which is located on the bank of Yanglin Lake in Songming County in the central part of Yunnan Province. As early as the early Ming Dynasty, the industry and commerce there have been prosperous, especially the liquor industry. Every year, when the autumn harvest came to an end, almost every family by the Yanglin Lake would make liquor, presenting a prosperous scene. Traditional wine making skills and rich knowledge of pharmacology are the solid foundation for the success of Yanglin Nutritious Liquor. At the end of the Qing Dynasty, Chen Ding, the founder of Yanglin Nutritious Liquor, set up a brewing workshop called "Yubao". Drawing lessons from the 18 water wine brewing techniques embodied in Lanmao's "Materia Medica of Southern Yunnan", ChenDing soaked more than 10 kinds of Chinese herbal medicines (such as Codonopsis pilosula, Hovenia acerba, Pericarpium Citri Reticulatae, dried Longan pulp, dates, etc.) in the self-brewed pure grain Xiaoqu liquor, and in the meantime, added appropriate amount of honey, cane sugar, pea sprouts, and green bamboo leaves. After a long-term exploration and practice, in the 6th year of Guangxu in the Qing Dynasty (AD 1880), ChenDing succeeded in composing the emerald green and limpid liquor which has a blended aroma of medicine and alcohol. This liquor is mellow and sweet, with a lasting aftertaste, and has the effects of invigorating the stomach, nourishing the spleen, reconciling the internal organs, promoting blood circulation and fitness. So the founder Chen Ding named it"Yanglin Nutritious Liquor".

Roxburgh Liquor

Both the Miao and Buyi ethnic minorities compose Roxburgh liquor. Different from the general compound liquor, water wine, rather than liquor, is used to soak Roxburgh. After having brewed glutinous rice wine, put dried Roxburgh in a cotton bag and soak it in the glutinous rice wine. Store the wine in the cellar for 3 months, then take out the cotton bag, and Roxburgh liquor is ready to serve. Roxburgh liquor is amber in color, delicious and mellow. It can invigorate the stomach, promote digestion and blood circulation.

Egg liquor

The egg liquor made by Yi people is a type of health-care compound liquor that is characteristic of ethnic culture. The preparation method is as follows.

1) Ingredients: Pure grain Shaojiu (vol. 40°-45°), ginger, Amomum tsao-ko, pepper, egg, sugar or brown sugar.

2) Boiling Shaojiu. First, Boil smashed Amomun tsao-ko, ginger, and sugar with Shaojiu. Then, scoop out the Amomun tsao-ko and ginger. Pour the beaten egg slowly into the boiled Shaojiu while stirring the Shaojiu quickly. At last, sprinkle some pepper, and enjoy the egg liquor.

The authentic egg liquor of Yi is always freshly cooked, so it is still warm when served, with tangy fragrance and refreshing taste. The egg white is silky, and the golden yolk is pleasing to the eye. It could relieve rheumatic pains. During the festive season, a bowl of hot egg liquor is full of jubilance and peace, expressing the sincerity and enthusiasm of Yi people to distinguished guests.

Songling Liquor

Songling liquor is a traditional beverage of Man people. Its production method is very unique. Man people would look for a pine tree in the mountains, and bury a crock of Shaojiu under the pine tree. A year later, dig it out to enjoy. It is said that this method enables the berried Shaojiu to absorb the essence of the pine tree. Songling liquor is amber in color and has the effects of improving eyesight and soothing the heart.

Chapter Three   Wine Vessels: Culture Bearers

"A good wine cannot taste nice without a good cup" is a folk proverb of the Yi people. Wine utensils are usually ethnic and they are an important part of ethnic culture. From primitive ceramic wine utensils to bronze wine utensils, bamboo and wood wine utensils, lacquered wine utensils, porcelain ones to ones made of precious materials such as jade, gold and silver, ivory, cloisonne, etc., various ethnic groups have their own features on their individual wine utensils.

Section 1   Simple and Durable Natural Wine Vessels

Natural wine vessels are made of natural materials. Some are made of animal horns, bones, hooves, claws, skins, shells, etc., and others are made of leaves, flowers, skins, and roots of plants. The former ones are horn cups, poultry claw cups, animal foot cups, leather wine sacs, parrot cups, etc.; the latter ones are lotus-leaf cups, grape-vine cups, birch wine utensils and those wine utensils made of gourds, coconut shells and bamboo tubes.

Horn cup is a wine cup made of animal horns. There are cow horn cups, goat horn cups, rhino horn cups and so on. Horn cups are still popular among some ethnic groups in the north and southwest of China. The Yi, Hani, Miao, and Wa people still retain the traditional custom of making wine cups with horns.

The beautiful and delicate horn cups of the Yi people are often decorated with lacquer or inlaid with jade. The horn cups are mostly used in pairs, and the tips of the corners are often drilled or knotted for carrying with ropes. In Liangshan, there are four classes of Yi people's horn cups, such as the first-class yak horn cups; the second-class Pianniu (an offspring of a bull and a yak) horn cups; the third-class ox or water buffalo horn cups; the fourth-class goat horn cups. The horns are boiled in water, the flesh inside the horns is taken out, the surface is scraped clean, and painted with lacquer. In Ailao mountainous area, Yi people's ox-horn wine and Miao people's croissant-horn wine are offered when guests come. Due to the round shape and the pointed bottom of the ox-horn cups and the croissant-horn cups, guests cannot put down the cups without drinking it up. In the Yi nationality wedding at the Walnut Valley in the Xishan District of Kunming, there is a custom of "Farewell-Guests Wine" with a cow horn cup or a goat horn cup. At the luncheon on the fourth day of the wedding, the bride and groom hold horn cups tied with a red thread and toast to the guests one by one. The guest must take the cup and drink it all, and then the couples make toasts again until the guest can no longer drink. Those who cannot drink leave the banquet immediately, and many people flee the banquet because they are unable to drink the wine up, or because they are afraid of losing their stance after being drunk.

The Yi people also use eagle claw wine cups or poultry claw cups, the upper part of which is made of bamboo, wood or leather bowls, and the lower part is made of poultry claws or animal feet. The foot cup is to remove the limb bones from the hoof of cattle and pigs, scrape the inner skin, put it into a wooden cup mold, and then dry it in the shade. The open hoof is used as the foot of the cup, and the mouth and body of the cup follow the shape of the animal foot, which is naturally inclined to one side.

The skin wine sacs are popular in many ethnic areas in the south and north of China. The whole sheepskin is kneaded, the head and limbs are removed. Leaving one leg as the mouth of the sac, the other perforations are tied and served; with wood and stone as wine cup model, the brewed cowhide or other thicker animal skins are tightly tightened on the model, and then the inner mold is taken out, polished and trimmed, and the surface can be painted into a patent leather wine bag.

Wooden bowl, wooden cup Tibetan, Mongolian, Yi, Menba, Hani, Nu, Lisu, Dulong, Jingpo, Jinuo, Achang people all make and use wooden wine utensils. They generally choose walnut wood, white gourd wood, catalpa wood, tsubaki wood and various chestnut wood with long years, fine and smooth wood pattern, and hard quality. According to their respective cultural habits and the capacity of the required wine utensils, the logs are intercepted and placed in a cool and dry place. After drying to the heart, they are peeled and hollowed out, then shaved, trimmed, and polished smoothly. Some wooden bowls and cups are made from roots of trees. If possible, they are painted with earthen vermilion lacquer, which is pleasing to the eyes. The clear liquor which is poured into the bowl is reflected in a bright amber color in the bottom of the cup.

Most of the Yi people's wine cups and pots are made of high-quality red chrysanthemum wood. The cups are mostly semicircular, connected up and down. The pots are mostly made of wood, flat and round. The wine is taken from the bottom with a thin bamboo tube, and people can use the bamboo tube to suck wine from the pot. There was a custom of using "up and down wine cups" in the Yi ethnic communities in Daliang Mountains in the southwest. The wooden wine cups are two ones of exactly the same size connected up and down. The bottom cup is completely painted black, and the upper cup is painted black and with colorful pictures on it.

Bamboo vessel and bamboo cups The ethnic groups grown up and developed in the sea of bamboo have an indissoluble bond with bamboo. Bamboo tubes are used for making cups to hold and drink wine, which is one of the important parts in the life of these ethnic groups, such as Yi, Lisu, Nu , Dulong,Gangba, Dai, Jingpo and Achang people.

Bamboo wine vessels integrate the function of holding and drinking wine. Served in bamboo tubes, the wine is permeated with the fragrance of the strands of bamboo and has a special flavor. The most common single-section bamboo wine vessel is to choose the mature bamboo in the bamboo forest, and intercept one-section according to the capacity, and keep the bamboo sections at both ends intact, then saw it into two sections, longer one as jug; shorter one serves as the lid, which slightly excavating the inner wall about one centimeter long, and the outer wall of the mouth of the jug being chiseled off one centimeter. When drinking, unscrew the lid and use it as a cup, and pour in fine wine to drink.

The bamboo vessels used by the Lisu and Nu peoples in western Yunnan is much more complicated. The vessel is made from naturally-bent bamboo in the mountains. People usually cut three sections into one part, the thicker end is used as a pot, leaving the bottom section intact; The upper bamboo joint is perforated to hold and pour wine; the thinner end is cut into an oblique mouth with a knife, which acts as the spout of the pot; as for the middle-curved bamboo joint , the two side walls are dug away, leaving only the top back as the handle of the bamboo vessel, which is natural, beautiful and durable.

There are two types of wine cups made of bamboo and its root. The bamboo cups are mostly made of hard golden bamboo in the mountains, take one section, cut into two, bamboo joints are the base, each bamboo section can be made into two cups. The bamboo cups used by the Dulong people are unique. They wrap the bamboo cups with binaural ears made of rattan, which are called binaural bamboo cups. Every time they drink, two people hold the same binaural cup, open their mouths and face to face to drink at the same time, which is called "drinking Tongxin wine." Many inhabitants in southern and western Yunnan such as Lisu, Nu and Dulong people can make and use bamboo root cups. They dug out the bamboo roots buried in the soil to dry, cut off the lateral roots, then hollowed out the main roots into containers, and polished the outside. A series of concentric circles are seen arranged on the outer wall of the wine glass in a staggered manner, making the whole wine glass look simple and lovely.

Many ethnic groups have the traditional custom of planting gourds in the courtyard. When the gourds mature, people hollow out the seed, tie a silk or rope on the thin stems, use the gourd to hold the wine when going out for farming, hunting, visiting friends and relatives. Because of its beautiful appearance, small size and large capacity, easy to carry and use, Yi, Miao, Lisu, Dai, Jingpo and other people like to make and use wine gourds. The Kucong people of the Lahu nationality not only use gourds to serve wine, but also toast with cut-open gourds. The traditional wine song goes like this: "Sweet wine, pour three bowls in a bamboo bowl; fragrant wine, pour three bowls in a wooden bowl; intoxicating wine, pour three scoops in a gourd scoop. Good wine for families and relatives, good meat for distinguished guests".

The main settlement of the Bai nationality is Dali, western Yunnan. The marble produced in Diancang mountain is also known as world-famous Diancang stone. The stone gives people a fresh and cool feeling because of its warm and delicate feature. The ancients called it "cold-water stone". Li Deyu, Xichuan governor, sobered after seeing marble when being drunk, so it is also called sober stone. The jugs, bowls, and cups made of marble are another wonder of the ethnic wine culture. In colorful marble wine utensils, we see sunset and twilight, mountains and tree shadows; in ink-washed marble wine utensils, we see distant mountains and clear waters, alternating between virtual and reality, fresh and pleasant; and the "Cangshan Snow Jade Cup" made of pure white marble is white, warm and radiant like jade.

Section 2   Ancient and Elegant Ceramic Wine Vessels

The use of fire not only enables humans to learn to cook food, but also urges them to further learn how to make pottery. The main pottery wine utensils include he, jia, pots and cups.

These early wine utensils were almost all used for both food and drinking or both for drinking water and wine, as well as for warming and mixing wine.

The ethnic groups have made pottery wine utensils in ancient times. The Kui pattern pottery of the Spring and Autumn Period and the Warring States Period unearthed from the site of Liujiang people, the ancestors of the Zhuang nationality in Wuming County. In the pottery wine jars of the Wa people, a bow-shaped bamboo tube with open bamboo joints is inserted in the middle when drinking. The other end of the bamboo tube is connected to a bamboo wine vessel. Using the siphon principle, the wine in the jar flows out for drinking.

Since the Ming and Qing Dynasties, the pottery casting industry has been developing day by day. Many pottery cellars are well-known for their exquisite, beautiful, practical and generous pottery, such as the earthen jars of Zhaozhou in Dali, the ones of Yongsheng County, and the clay pots of Kunming Clay Kiln in western Yunnan are all excellent wine containers. Among the drinking utensils, the five-color pottery of Jianshui in southern Yunnan is "exquisitely made, including everything that one expects to find". With its feature of "sound like chime stone, clear like water, and bright like a mirror", Jianshui pottery has also become the top one of wine utensils.

The Tibetan wine-making utensil "PengZan" is pottery too. According to legend, it was introduced to Tibet by Princess Wencheng of Tang Dynasty. The height is approximately 60cm, the bottom of the retort is used to hold water; the middle part is used to hold highland barley; the top of the utensil is placed in a sky pot with cold water; there is a funnel under the sky pot. The grain has been cooked for a long time and distilled into liquor through a funnel.

Section 3　Elegant and Noble Metal Wine Vessels

Gold and silver wine vessels are not owned by ordinary people. Most are used by ethnic aristocrats and chieftains, which is a sign of status and mostly used for rituals, large-scale banquets and other activities. The Mongolian, Yi, Naxi and other ethnic groups use gold and silver wine utensils to express their piety when they hold a ceremony to worship the sky. Ordinary people also occasionally collect gold and silver wine vessels, but they only use them in important events such as ancestor worship. Mosuo people in West Yunnan sing to worship their ancestors: "The silver bowl is filled with wine, please open your eyes and have a look, drinking a bit will heal foot pain; the golden bowl is filled with tea, please open your eyes and have a look, tasting a bit will cure your pain".

The bronze culture of ethnic groups is relatively developed, so copper wine utensils have a long history, and many of them have been discovered in the unearthed cultural relics. Since the Ming and Qing dynasties, people have used copper wine and tea sets. Among them, the spotted copper and red copper ones from Dongchuan, Yunnan are the most distinctive, practical and beautiful.

Gejiu in Hani and Yi Autonomous Prefecture of Yunnan Red River is the famous Tin capital. Tin crafts have a history of thousands of years. Since the late Qing dynasty, Yunnan tin wine utensils have been used at the banquet and ordinary people's homes. Tin wine utensils are crystal clear, elegant and beautiful in shape, moisture-proof and good at heat preservation, acid and alkali resistant.

Section 4　Bright and Elegant Lacquered Wine Vessels

The origin and development of lacquered wood wine utensils in our country can be traced back to the Hemudu period about 7000 years ago. At that time, the lacquer bowls were mainly used as food utensils, not special wine utensils. In the Xia and Shang dynasties, although lacquer painting technology had been rapidly developed, but at that time, people mainly used bronze ones. In the Spring and Autumn Period and Warring States Period, lacquer wood wine utensils had a great development, such as lacquer spoons and wine vessel boxes unearthed from the tomb of Chu. However, lacquered wood wine utensils really replaced bronze wine utensils only in the Qin Dynasty, and its prevalence was in the Han, Wei and Jin Dynasties. Most of lacquered wood wine utensils are painted with black lacquer outside and vermilion lacquer inside, and the pattern with vermilion lacquer, which looks solemn and elegant, beautiful and generous. Lacquered wood wine utensils are also moisture-proof, anti-corrosion, easy to clean and light. Today Yi people's wine utensils are mostly decorated and painted with lacquer, such as the horn and wood cups. The wine utensils of the Yi people in Liangshan are mostly wood, with purple-red paint as the background, painted with black and yellow patterns, which are very elegant and unique in style. One of wine jars is called "Salipao" (or "Sanabao") in Yi language. It is equipped with a straw and made of wood, round and flat like a drum, with a straw on the shoulder. With an empty tube in the center of the pot, wine is injected from the bottom and flows into the pot. When drinking, the wine is sucked from the straw into the mouth.

Chapter Four   Wine Etiquette in Long History

China has been known as the "state of etiquette" since ancient times, and etiquette occupies a very important position in Chinese social and cultural life. There are many ceremonies in a person's life, such as birth, full moon, hundred-day, one-year-old, adulthood, marriage, funeral, etc. Among the ethnic groups in Yunnan, rituals and wine are also linked together.

Section 1   Wine and Birth Etiquette

Childbirth is a major and happy event for a family or even a clan. Certain ceremonies are held to celebrate, and to inform the family, neighbors and friends. Wine has become a must-have in this occasion.

Yi people take pregnancy and childbearing as joyful events. After the child was born, the son-in-law went to his father-in-law's house to report the good news and brought a bottle of wine and a chicken. If a boy, brought a hen; if a girl, brought a rooster. The father-in-law took the hen and exchanged a rooster to the son-in-law; or the father-in-law took the rooster and exchanged a hen. The son-in-law brought back the changed chickens and raised them, and he couldn't kill them. After hearing the good news, the mother-in-law prepared a jar of boiled liquor, a baby's back, a cloth, a set of clothes, and hundreds of eggs. Relatives and friends of women came to send eggs during the month of childbirth. When the child is full moon, these relatives and friends should be invited back to eat and drink for a day.

When the child is at the full moon ceremony, parents will ask Bimo to recite the scriptures and ask the elders to shave his hair. If the child wants to keep long hair, just shave until the child is three years old. In Yi Village adjacent to Yunnan and Guizhou, there is a custom of "hanging red" when the child is full moon. Relatives and buy and bring the cloth, and hang it on the door of the house where the child is born. The door is nailed with a number of nails, and the cloth is hung from above to the ground. If a child often cries and is upset before one year old, parents need to take the child to the road and ask for a name. The parents prepare a bottle of wine, a cooked chicken, a pot of rice, make a small wooden bridge, and put these things on the small wooden bridge beside the roads to mountain. The child and the parents hide and observe nearby. Seeing a twenty-year-old man crossing the bridge, they ran out and grabbed him, grabbed a button, brought the child forward and asked for a name. When the person named the child, he carried the child and worshiped three times in the south, east, north, and west. The young man became the god-father of the child and they became the relatives. The two sides cooked hot chicken and rice on the spot, toasted and drank wine, and told each other their names and places of residence when parting, and they would visit each other often in the future.

To shave long hair of three years, the parents must treat once. Before they ask an elder to shave child's head, Bimo must recite the "Long Hair Sutra": What day is today? It's a good day. You ask me to come today, and I will recite the scriptures today. What am I reciting? I'm reciting the shaving sutra. The dad listens, the mom listens, the child listens with ears up, relatives and friends listen carefully. The child, you know, you are the bones of Abba, you are the flesh of Ama. At the full moon, keep you long hair and treat you as a baby. In a blink of an eye, three months are full, three months and ten days, a full hundred days, you will laugh; in a blink of an eye, six months and six days, you will have teeth, and you will laugh. In the blink of an eye, with many days and nights, you can sit after seven months, you can climb after eight months, and you can stand up and stagger after nine months. Mom fears that you will fall, Abba fears that you will fall, watching on you until one year old. You can walk and you can run. You eat when hungry, and you drink when thirsty. When you are two years old, you call mom and dad. You can talk at the age of two. Dad calls you baby and tells you to take a cigarette pot; Mom calls you baby and tells you to pick up the thread balls. Now you are three years old, shave your long hair, and learn to be polite. When you are seven years old, we will send you to the school and receive the teacher's training. Child is the father of the man. We hope you will grow into a talent, so that one day you can be a pillar for the country.

In Yongning area, a sun worship ceremony will be held on the third day after the child is born. As soon as the sun came out, the mother or sister of the parturient threw a burning pine in the yard. The mother held the baby in her left hand, and a mirror knife, a hemp pole and a page of Daba Sutra in her right hand. She sat in the patio for a while, soaked the baby in the sun, and prayed for sun blessing. The mother of the parturient cooks Sulima wine, pig fat and various staple foods to entertain the older generation of women in the village.

On the seventh day (boy) or ninth day (girl) after Dulong babies' birth, a naming ceremony is performed. Family members and neighbors of both parents go to congratulate, and babies are named by the child's father or well-known elders in the clan. The master will slaughter chickens and pigs, and make rice wine to entertain relatives and neighbors.

The Miao nationality has a custom of children finding a guardian. Usually based on the children's eight characters, parents will look for elder one with harmonious eight characters to protect their children. If the child's Bazi (eight characters) belongs to water, the guardian would be someone whose Bazi belongs to the soil to preserve water with the soil. Usually on an auspicious day, the guardian and the relatives of the family will be invited to have a reunion drink at home. During the banquet, an elder relative will explain that a child respects someone as his or her guardian. The wine drunk during the ceremony was brought by relatives as a blessing. Then the child will go to the guardian's house to stay for three days, he will give a silver neck-circle to the child, which is called a life-saving ring. Three days later, the child returns to his home to worship his ancestors with wine and meat, and since then called his parents uncle and aunt.

Section 2　Wine and Adult Etiquette

The coming-of-age ceremony is a ritual for a person to change from adolescence to adulthood. In ancient times, Han people call it the crown ceremony, and most men were crowned in their 20s. A branch of the Miao nationality, the coming-of-age ceremony of the antique Miao men is called getting the old name or the honorable name. On the second day of their marriage, the old and young of the Jufang clan held a grand naming ceremony. Several elders in the clan first secretly agreed on a name, and then set up a banquet in the main room, and invited the brother-in-law to sit on the important position and guessed the name until getting the correct answer. When the brother-in-law stepped out of the gate to announce the name, those waiting outside must offer three bowls of wine to him, and there is a rule that anyone who called the wrong old name would be fined 360 catties of wine and 360 catties of meat! After the announcement, three bowls of wine and some congratulations would be offered.

The hair shave ceremony of the Yi nationality is also a coming-of-age ceremony and requires drinking. Children of the Yi nationality all grow long hair since their childhood, and when they grow up to seven or eight years old, their parents must calculate the date of shave based on their birthday. The long hair must be shaved by the child's uncle, and the married and engaged sisters and aunts must prepare gifts for drinking wine.

Section 3   Wine and Marriage Etiquette

At every step in the process of the marriage of all ethnic groups, from making friends and falling in love with young men and women, proposing marriage, getting engaged, giving dowry, welcoming bride and holding a wedding, wine plays an important role.

Spouse choice: fair lady, good gentleman's choice

Young men and women of ethnic groups generally have greater freedom in choosing a spouse. Festival drinking, singing and dancing are often the best opportunities for men and women to choose a spouse, such as the Yi's Torch Festival, Miao's Huashan Festival, Bai's March Street and Raoshanling Festival and other traditional festivals or activities. Some ethnic groups have special places for young men and women to meet and date, such as the "public houses" of the Dai and Hani peoples, and the "girls' rooms" of the Miao people in Chuxiong. Young men and women in different places usually date in the mountains during the slack time after the autumn harvest and planting. Young people drink, eat sugars, tease and laugh, observing and choosing the person they like in cheerful, antiphonal dances.

The Yi people in the Chuxiong mountainous area and Luquan area have the custom of "eating mountain wine". During the Spring Festival, the Yi boys and girls carried sweets and wine, cheering and walking towards the beautiful and secluded mountains and wilderness, burning a bonfire, and then opened the prelude to mutual understanding in the form of antithetical songs. The two parties improvised and responded to the songs, approaching the bonfire, eating sweets, drinking wine, making fun of each other, stepping on the cheerful bamboo flute syllables and dancing. Liquor has a very important "catalysis" effect in expressing emotions. There is a saying of "thirty percentage of liquor and seventy percentage of courage". On the fifteenth day of the first month of the lunar calendar, the Yi people in Dayao and Yongren counties of Yunnan kill chickens and goats, drink and sing in antiphonal style. The girls wear their own elaborately embroidered costumes to celebrate the annual clothing festival.

Proposal and Engagement: Sign a Happiness Contract

Eating and drinking in the woods, singing together in the fields, the men and women are falling in love with each other. After this initial stage of marriage, the subsequent proposal and engagement will enter the formal marriage process.

After the young people in the Keno Mountains fall in love, the man can spend the night at the woman's house, but he will leave early in the morning. The early stage of this marriage can last for 1-3 years before getting engaged and married. Engagement in Batu and other villages is divided into three stages: "Yinei", "Axiaani", and "Ajupiao". In the first stage, the man invited three or five relatives and the head of the village to drink for three consecutive nights; the next one or two months enter the second stage, drinking for one night, mainly for both the man and the woman to count eight characters; the third time is a formal engagement. The man brings a few coins and twenty or thirty bamboo barrels of liquor, and asks the woman to give it to his uncle. After both parties drink for one night, the engagement ceremony is basically over.

The proposal and engagement of the Achang ethnic group in Lianghe County, western Yunnan, is also quite distinctive. After giving the invitation wine, engagement wine and betrothal wine, the bridegroom will also give the bride eggnog. The man's side selects 6 fresh eggs, rice wine, let the matchmaker bring to the woman's house. The man breaks the eggs one by one in front of the woman's parents, relatives and friends, put them in a clay bowl, add some rice wine and stir evenly, makes a toast to people in the bride's house.

As for the Naxi people in the Sanba area of Zhongdian County, Yunnan, the matchmaker is the man's uncle, he only brings a can of wine and a piece of sugar when he goes to the female's house. When he comes to the girl's house, he opens the door and explains his intentions, and then starts to fetch water and wash hands, later removes the censer from the cabinet, burns the incense and kowtows. After the ceremony, the girl's parents said: "When we got married in the sky, my uncle brought a white horse with a golden saddle; when we got married on the ground, my uncle brought a black cow with a wooden pen. Uncle came here today, what did you bring with you? The uncle replied: I only bring sweet brown sugar and fragrant rice wine. With his real heart, the man will make your daughter feel warm and happy"! After the question and answer, if the woman's parents agree to the proposal, they will leave the matchmaker for tea and dinner, and the matchmaker will take out the hidden rice wine and give it to the girl's parents.

After the engagement, Wa man will give a "dupa" gift of three wines to the woman's family. The first time is to give "Bailaire", namely clan wine: 6 bottles of liquor, plantain, tea, etc., and it will invite the male head family of the father's clan to drink together; the second time is to give "Bailaimeng", namely neighbor wine, for 6 bottles of liquor, this time will invite neighbors to come and drink as an eyewitness of the marriage; for the third time, it will send "Bailaibaoxiwai", that is, open-the-door wine. The girl's mother will leave a bottle of liquor by the pillow, drink quietly at night and pray to the gods for her daughter and wish her happiness in the future.

Examining the marriage proposal and engagement customs of various ethnic groups, we can find that wine is almost an indispensable gift, and they are all "small pot wine" and " distilled pot wine" brewed by their own family.

Welcoming bride (go to bride's home to escort her back to wedding): Drinking and Singing Antithetical Songs

The basic procedure of a wedding is composed of a trilogy of the man coming to meet and welcome his bride, the families and relatives send the bride away, and the two sides hold the wedding ceremony. Although the ceremony is slightly different from region to region, "wine" is everywhere.

The welcoming ceremony is often held both at the woman's house and the man's house. The Yi people send a team to the bride's home, and after the bride puts on the clothes, ear rings and bracelets sent by the mother-in-law, people eat "Ditang wine." Ditang wine is a kind of ceremonial drinking. The woman's family invites relatives and friends to sit in the main room. Three people, the man, the woman who send the bride away, and the bride will turn around three times in the main room from right to left and then give everyone a toast in turn. The night feast is the liveliest scene of the wedding in the woman's house, and it is also the period when the atmosphere reaches its climax. The format is as follows: Two tables are placed in the main room, and 12 people are seated. There are fourteen cups, three bottles of wine, two bowls of rice, and two spoons on the table. Mr. Wedding greeters buckle twelve cups on the table and the remaining two cups are put on a bottle of wine and then put under the table, then he puts the cups on the table upright again, pours the wine, and then begins to tell the story.

The Yi group returns to the man's house the next day, and relatives and friends come to congratulate him with noisy firecrackers and joyful atmosphere reaching a climax. After lunch on the second day of the wedding, all relatives and friends gathered together to drink sweet wine for congratulations and sang a toast song. The men of the family brought a bottle of wine to the welcoming team's residence to drink and have fun together. After the bride enters the bridal chamber, while making the bed inside, the brothers outside put a big bench at the gate, and place two jars of wine on it, with four to eight sticks inserted in the jar (12 ones for three generations). The sticks are tied with flowers. People hold the stick tightly with their hands, and when listening to the "please" from the wine officer, and the slickers immediately start to drink. The wine officer said "stop" and everyone immediately stops drinking. If the person does not stop or does not finish the prescribed amount, he shall be punished and drinks again.

The Shui people express the commitment of both parties to their marriage by drinking, which is called eating big wine. When picking up the bride, the man's side usually sends two men and two women, and among them one group must be a couple. They both represent the man to greet the bride, and they are also witnesses of the marriage. When eating big wine, a long table is placed in the middle of the hall, the pig's head in the middle of the table, and three sea bowls on each side, one of which is full of wine, two of which are half bowl of wine, and money is pressed underneath. Each party will elect three highly respected senior citizens who are good at mediating disputes to come to the feast. After drinking a few cups of good old wine, the drinkers on both sides began to question and answer each other.

Then the two drank the "big wine" in one go. Then one person from the woman's family accompanies the man's family to drink the first half bowl of wine, which means: in the future my daughter comes to your home, if someone tells gossip, I will ask you to be responsible for mediating and calming down. Another person from the woman's family accompanies the man's family to drink the second half-bowl of wine, which means: you are responsible for my daughter when we send her away. At this time, there was thunderous cheers in the audience, and everyone drank the "communication wine" which indicates that the relatives of both sides would always be friendly. Then they had dinner together, drank and sang for congratulations.

Wedding: the newlyweds drink hands-over wine, friends and relatives swig

The wedding is the most important part of the socially agreed marriage procedure, and all ethnic groups have their own distinctive drinking customs and rules in the wedding.

Marriage is a top priority for Hani people, and they will invite their relatives and friends to have a wedding drink. At the wedding, the bridegroom carried a wine jug, and the bride carried a wine cup, and bowed to toast the guests in turn. The guests say some blessings after receiving the wedding wine, and then put the silver gift they have prepared on the newlyweds' plate, and finally a toast. The bride will eat the half-raw rice handed over by the groom, indicating that she is married to the groom. Being willing to eat the half-raw meal as cooked rice, the bride will never change her heart to the groom. The bride also gives the groom the round cakes brought from her natal family, expressing her willingness to reunite with the groom and never be separated.

The most distinctive custom in Pumi weddings is "taking the key" and "locking the matchmaker". When welcoming the bride, the matchmaker gives gifts to the girl's house. After the bride going on the road, the matchmaker and a singer selected by the girl's side are locked in a room, and another girl guarding the door. The matchmaker and the girl singer will sing duet in the room. If the matchmaker wins, the girl guarding the door opens the door and let her go; if she loses, she must drink a sip of wine from every family member in the whole village. There are often drunk matchmakers surrounded, laughed at by the bride's friends, brought home with four feet upside down. There is no doubt that the Pumi matchmaker must have a mouth that can sing good songs and drink good wine. When a girl gets married, a group of people form a sending-away team, and the key to the bride's dowry box is kept by an elder among the relatives. When he arrives at the groom's house, the man has to conduct the custom of "getting the key" to open the cabinet. The groom's family appoints a representative to hold the tray, symbolically put a few coins and a bottle of wine on the tray, and standing respectfully in front of the relatives, asking for the key by singing. After the people who sending the bride sing back , accept the wine and money, and then put the key in the plate.

In Jingdong, for Heiyi people, the bride and groom both enter the bridal chamber to "grab the headboard", whoever grabs the headboard and sits down is smart. After sitting down, a "spraying wine" ceremony was held. The groom's close friends and peers pour 2 glasses of wine and hand them to the bride and groom. The newlyweds make a toast, leaving a little wine in their mouths, and sprayed to each other's face. Whoever sprayed it first wins. After the "spraying-wine ceremony", there was a "falling-on-frame ceremony". The bride and groom drink a cup of wine again, and hurry to get out of the bridal chamber together, vying to step on the threshold, which is the so-called "falling-on- frame". It is said that whoever steps on the threshold and walks out of the bridal chamber first will have more power in the family in the future.

Drinking hands-over wine in the wedding of the Bai people in Dali was hosted by a woman with many children and grandchildren. Two identical wine cups were prepared in advance and tied together with a five-color thread. The bride and groom sit on the wedding bed and drink with their arms crossing together. The host sings: "Pour full wine, and raise your son to be a champion; Pour full wine and raise your son to be a teacher" and so on. At the Yi wedding in the Ailao mountainous area, the bride and groom use two scoops of a gourd to hold the wine and drink. After drinking, the two scoops are combined into a complete gourd. According to the investigation by Mr. Liu Yaohan, an expert on Yi culture, this ancient ceremony symbolizes that the newlyweds have also become a combined gourd, that is, that means the couple combine with the same respect and inferiority, and their souls will also enter the same gourd after death.

Yao people drink "Lianxin Wine" when they hold weddings. On the wedding night, the man's family feasted the guests. The bride and groom, matchmakers, parents, brothers and sisters, younger siblings, main relatives and friends, patriarchs, and foreign guests were seated at the first big table formed by 5 tables. After the bride and groom filled each guest's cups with wine, then the wine is poured back into the jug and mixed together, and then poured into each person's cups. This wine is called "Lianxin Wine". When the host of the wedding party shouted "Cheers", the "Lianxin Wine" feast began. The bride and groom make a toast to every elder, relatives and friends. Every time they make a toast, they will drink a cup by themselves.

On the weddings of Man nationality, people will offer sacrifices to gods at noon. There is a god table in the courtyard to put pork knuckles, three wines on a plate, and a sharp knife. The bride and groom kneel at the table facing the south, and the Saman kneels on the left side of the table and sings "a song of Achabumi". The sacrificial song is divided into three sections. After each section, he cuts a piece of meat with a sharp knife and throws it into the air, taking the wine for a while, raise it to the eyebrow, and then pour it on the ground. At the end of the song, the drums rang together, pushing the wedding to a climax, which is called "Sazhan". In the evening, the couple will eat the hands-over wine and go through the marriage ceremony in the bridal chamber. A table is set up in the bridal chamber, and the bride and groom walk around the table hand in hand three times. People with good fortune fill up two cups of wine. The couple sip each, then exchange wine cups and drink another sip. Then they eat the children steamed-buns and longevity noodles, and then the two sides fight to sit on the bed, thinking it is a good omen. At that time, the bride and bridegroom compete to sit on the quilt. As a result, the two of them often sat on the quilt at the same time. On the wedding night, the candles in the bridal chamber will not be extinguished all night, and there are people outside who sing a joyous song called "Xiangfang". Someone sprinkles soybeans to the bridal chamber, wishing that the couple would be more prosperous and give birth to an early son after marriage.

Section 4   Wine and Funeral Rituals Etiquette

Funeral is also one of life rituals, and wine plays a certain role in all aspects of funerals.

Bring wine to inform the death to relatives

After death, people need to report death to relatives and friends. Many ethnic groups have the custom of carrying liquor to report the death. The Buyi people on both sides of the Red River fire three muskets and beat drums to report to the village about the death. Then the children or nephews of the deceased go to the door to bow down and present a pot (bottle) of wine and a white fidelity handkerchief. The Yi people in the Shao area of Chuxiong County put a small piece of white linen cloth under their hats, brought a chicken and a pot of wine to the house, stopped in front of the hall, threw the chicken into the hall, and placed the wine in front of the door, then bowed down outside the door. After the family caught the chicken and collected wine to prepare for the funeral, the person even didn't say a word in the whole process of reporting the death. All ethnic groups generally believe that reporting death will bring filthy evil spirits. After the person left, the people who are notified should take a sip of wine and spray the wine in the house and use a broom to wipe it out in order to get rid of the evil spirits. Those who rich ones even ask the priest to recite the scriptures to drive evil spirits away after the mourners left.

Bring wine to funeral

On the day of the funeral, the Sani people in the Guishan area of Lunan, Yunnan, all men, women and children in the whole village bring wine and vegetables to the funeral. The man walks in front of the coffin and the woman walks behind the coffin and all people go straight to the cemetery. After the coffin is put in the soil, they drink and eat dishes to comfort the bereaved family members, then came back to the village together.

Among the items carried in the funeral, the most important thing is wine. The amount of wine varies, depending on the relationship between the bereaved and the mourners, as much as tens of catties and as little as one or two catties. In Wuding and Luquan Yi ethnic communities in Yunnan, distant relatives carry one catty of wine and a chicken for funerals; if close relatives, they have to carry the sheep and wine. The Mosuo people in Ninghen Yi Autonomous County of west of Yunnan, relatives and friends only need a bottle of wine and a few tea for the funeral, their children need to bring cattle, sheep and cans of wine for funerals; if the deceased is a female elder, the uncle carry the most gifts to the funeral. After the mourners put the wine on the altar table in front of the coffin, they are taken by the chief mourner. After the funeral, the Mosuo people have the custom of "returning the gifts". Sometimes an ox leg and an ox head, and the person who carried the wine leaves a little wine at the bottom of the bottle or can, and declared that the gift was returned by the deceased and it was brought home by the mourners to worship the ancestors.

Use good wine to sacrifice the souls of ancestors

Many ethnic groups believe that people are dead but souls are still alive. Therefore, there are a series of aftermath activities after the corpse is put into the grave. The Yi people believe that people have three souls, one soul returns to the ancient residence of their ancestors after death; one soul stays at the cemetery to protect the offerings; and another soul accepts worship at home to bless the family's prosperity. If the latter two souls are not conscientiously worshiped, they will become wild ghosts and endanger the living ones. After the corpse is buried, there are activities such as collecting souls, making ancestral cards, and chanting spells to release souls of the dead from suffering, which are a continuation of the funeral activities.

The Pumi people regard death as life. During the morgue period, the spirits of the deceased continue to be served. Before the coffin is put into the earth, good wine must still be offered so that the deceased can carry the wine and return to the heaven. The Pumi Elegy (Diaoeh Tune) in Tuodian, Ninglang Yi Autonomous County, western Yunnan, sang:

Before you leave, I will offer you fine wine, this bowl of fragrant fine wine comes from hard sweat. This bowl of intoxicating wine is dedicated by the four families of Pumi.

The eldest son sat on the top, a Tibetan, I asked him for highland barley.

The second son sat in the middle, a Mosuo, I asked him for bitter buckwheat.

The third son sat downward, a Pumi, and I asked him for barley.

The fourth son sat at the end of the table, a Han, I asked him for rice.

The shepherd in the high mountains sent bitter distiller's yeast. The pig-keeper on the river side brought sweet distiller's yeast. The sheep herders in the village brought sour distiller's yeast. The bitter, sweet and sour yeast make the wine, the sweetest Sulima wine.

Today, I put Sulima in the golden horn and silver horn cups, holding it in front of you with both hands. If you look at it with your eyes, your eyes will be brighter; if you taste it with your tongue, you will speak faster; if you take it with your hands, you will succeed in everything; you can go anywhere with a drop on your foot.

Your head will become a conch tree, your eyes will become sun and moon, your teeth will become nebulae, your tongue will become a rainbow, your blood will become the sea, and your bones will become mountains, your intestines will turn into roads, and your limbs will stretch into four directions. You use one hand to hold the dedicated sandalwood tree cup; you use one hand to hold the dedicated wine bowl, to pursue your happiness, to find your joy!

The ghosts bring good wine

Among the rich and colorful funeral customs of ethnic groups, there is also a funeral activity called "sending wine." The so-called sending wine means that the living asks the souls of the deceased to bring fine wines for their departed relatives. This custom is still kept intact among the Pumi and Mosuo people in the west Yunnan.

On the beautiful Lugu Lake, Mosuo people not only like to drink, but also believe that the dead also like to drink. Since the living cannot communicate with the dead, they have the custom of asking the person who has just passed away to bring a good wine for their long-dead relatives and ancestors. The Mosuo people call the day of concentrated funeral "Buzu". On this day, in addition to their relatives and friends who came to the funeral with wine, Mosuo people of different clans from the neighboring villages also brought wine and offered it in front of the coffin. The dead were expected to bring the wine and see their gone relatives and clan ancestors. The sender must hold the wine bowl in both hands, raise to his head high, kneel in front of the deceased's spirit, and have the scripture master sing the song of sending wine before the entrusting was completed. The sending-wine songs go like this:

Not long ago, Pumi people passing by, and I wanted to ask him to send the wine to my ancestors, but it's impossible because I didn't understand Pumi language.

Han people coming the next day, and I wanted to ask him to send wine to my ancestors, but it's impossible because I didn't understand Han language.

One day a forgetful horse driver passing by, I dare not ask him to send the wine because I was afraid that he would forget the wine on the road.

One day the wild goose flying over its head, and the forgetful wild goose feared that it would spill the wine among the white clouds, and I dare not ask him to send the wine.

One day a wild duck flying over the village, and the forgetful wild duck was afraid that it would pour the wine on the river and grass beach, and I dare not ask it to send the wine.

In our village, you will see our ancestors soon. I don't hesitate to bother you to bring messages. Oral messages are like breeze, not difficult to bring, the wind will help you bring them there.

Chapter Five   Unique Wine Customs

Make friends with wine, develop friendship with wine, and narrate love with wine. Hospitality with wine are almost the common characteristics of ethnic groups. For the Yi people in Luquan area, when their relatives and friends leave, they will sing a song to retain the guests, declaring that "there is a lot of rice wine, and after drinking nine small jars, there are still ninety-nine ones."

Section 1   Warm and Passionate Hospitality Customs

The important role of wine in interpersonal communication is mainly used for treating diners. Wine is not only used to show welcome and respect to guests, but also to express and promote the joy of the host and guest gathering. People of all ethnic groups attach great importance to serving guests with wine. The Yi proverb says: "Han people cherish tea, and Yi people cherish wine."

In the Miao nationality area, ox-horn wine is a famous and noble hospitality wine. Because of pointed bottom of the horn cups, the cups cannot be put down, and guests must drink wine all at once. Since ancient times, the Miao family have regarded cows as treasures and made friends with them. When cows give birth to calves, people feed them soy milk and glutinous rice porridge; on New Year's holidays, the cows are fed some meat, fish, eggs, and some wine with bamboo jar to reward the cows for their hard work and people will share joy with the cows. After the cattle died, the Miao family was very distressed. To remember the cow, they saw off the horns to make wine cups and hung them in the house. When guests enter the house, they need to drink one horn-cup "door wine". The guests are seated and offered with three cups of wine, and then they can drink "turning-wine" and "hands-over wine" together. During the holidays or festivals, or when a distinguished guest arrives, people will drink and toast with horn cups, which not only expresses respect and love for the guests, but also shows the nostalgia for the cow. Some people often hang a pair of horn cups in the wooden building at the gate of the village. When the distinguished guests enter the village, the old villagers or girls in ancient costumes hold the horn cups in both hands and toast the guests one by one. It is Miao's most noble welcoming etiquette.

Farewell wine is etiquette of Miao people. The guests are leaving, and the host is reluctant to part with them and they send them away with wine at parting. The host and guests each hold a bowl full of wine, and the host sings and says goodbye to the guests by drinking wine. If the distinguished guests, people are even more passionate. They first stepped on the bronze drum and performed the reed dance, and summoned all villagers to say goodbye to the distinguished guests. Men, women, young and old went down the wooden building one after another, came to the Tonggu Grassland, and danced around the bronze drum. With a table of several bowls of wine and a few pieces of red silk on it, the old villager announced to see off the guests, then he hung the red silk diagonally on the guests, and suddenly there was joy from the crowd. People came up one by one, sang a farewell song, offered a sip of wine, and tied colorful flower ribbons to the red silk. When the guests walked out of the village, the host persuaded the guests to take another drink before finally parting.

When Tibetans serving guests with highland barley wine, they first fill the cup with wine and bring it to the guests. The guest should take it with both hands, and then hold the cup in one hand, and put the middle finger and thumb of the other hand into the cup lightly, and flick it toward the sky, which means worshiping the gods; then repeating it for two or three times to respect the Earth and the Buddhas. When drinking, the established custom is to take one sip, and the host immediately pours the wine to fill the cup; then takes a second sip and fills it again; then drinks a third sip and then fills it up. After that, you have to drink a full cup of wine in one sip. In doing so, the host feels that the guests value him. The more the guests drink, the happier the host is, which shows that the host's wine is well brewed. When the Tibetan people toast, they use large cups or bowls for male guests, and small cups or bowls for female guests. Being good at singing and dancing, Tibetans inevitably sing wine songs and perform "Guo Zhuang" dance (a folk dance of Zang nationality) when drinking.

The Mongolian people have a lot of hospitality etiquette, including offering cigarettes, toasting wine, offering Hada Deji, giving whole sheep and whole cows. The host makes a toast, and the guests take it with both hands. Guests are not expected to hand things to the host with their left hand. At the banquet, whether you receive a wine or a toast, you must pull down your sleeves. Pouring wine and toasting guests is one of the most common traditional etiquette for the Mongolian people to entertain guests. Mongolians believe that fine wine is the essence of food and the crystallization of grains, and offering fine wine to guests can express the respect and love of the pastoral shepherds.

Buyi people are polite and hospitable. When distinguished guests arrive, there must be six wine rituals, including door-wine, hands-over wine, Biandang wine, turning-wine, a thousand-cup wine, and seeing-off wine. If offering pork, it is to wish the guests to raise pigs well and a good harvest in the coming year; if offering chicken, the head is presented to the first guest, symbolizing good luck; wings are presented to the second guest, which means taking off successfully; and legs are presented to the third guests, which means down to earth. Toast Song and Night-meal Song are sung on the banquet. The former song is to persuade guest to drink the wine, and the latter one is to sing all the items and food on the table one by one to show their minds and talents.

When Buyi people toast, they toast three bowls of "Biandang wine", known as "three-turn wine." The guests have to drink a bowl every time. If they want to escape, the girls will sing toast songs constantly, sincere and passionate, guests are difficult to refuse. When treating guests in the village, the host family entertains them on the first day, and the whole village takes turns to entertain guests from the next day. This custom of eating "Hundred-Family meals" reflects the tradition of the Buyi people's hospitality. Every household brew rice wine and drinks it at home during the slack farming period. This type of wine is easy to make, low degree and with a nice taste, sweet and delicious. This wine is used to treat the guests on weddings, funerals, house-building ceremony, full-moon ceremony, and birthdays.

The warm and hospitable Zhuang People have the custom of toasting 12 cups of wine to the guests. Number "12" is closely related with traditional culture of Zhuang nationality. In the early Zhuang mythology, there were 12 suns, 11 of which were shot by LangJin; a year is divided into 12 months, and a day is divided into 12 hours; the zodiac sign of a person is set as 12 ones; a living person has 12 souls; when getting married, the man needs to give the woman 12 silver coins, 12 catties of wine, and 12 large glutinous rice cakes. The number of people to send the bride is 12 people; they offer 12 cups of wine to the gods; people who send the souls of the dead to the ancestors must cross 12 rivers and 12 bridges, climb 12 mountains, and pass by 12 villages. Therefore, there is a custom of toasting 12 cups of wine to guests. When drinking handover wine, they don't use a cup, but a white porcelain spoon. The two people scoop a spoonful one from the wine bowl and drink with each other. The host will sing a song:"please drink the wine in the tin jug presenting to you. I show my respect to your distinguished guests with a sincere heart. Please drink wine in white porcelain cups, though it isn't very good wine but with my heart. When toasting, you are just like a god in my heart.

Hani people have the custom of drinking stewed pot wine. After guests enter the house, they are seated on the upper table by the fire-pit and offered a bowl of fragrant rice wine, which is called "drinking stewed pot wine". The hospitable Hani people prepare a rich meal for the guests, and they are invited to take a seat at the upper table. They fill the guests' cups first, and then pour other people's cups. Before the toast, the host lightly said the blessing words, then he used his index finger to take a little wine from his cup, and marked the Chinese Character "one" on the table in front of him and on his forehead respectively to express exorcism and wish all people happiness and health. Everyone toasted and drank. Juniors cannot accompany guests when drinking, and the owner cannot sway his legs. When sending guests off, they put some gifts in the pockets of the guests, such as eggs and rice balls. It is considered impolite to let a guest go back empty-handed. Hani people usually only toast three ones, because in their eyes, three is a very beautiful and auspicious number. The most important thing is that Hani people think that three cups of wine are the most suitable and helpful to one's health, and drinking too much will harm our body. After drinking three wines, they will not toast anymore. If guests want to drink, they can drink more. If they don't want to drink, they can eat. People will never deliberately get their guests drunk. They just wish them happy and healthy.

There is a saying in Pumi people: "The first bowl of highland barley wine poured out of the new jar must be toasted to visitors from afar; the first cup of tea from the newly heated tea pot must be given to the distant brothers."

The Nu People still keeps a unique custom of welcoming guests. After the guests' arrival, the host first takes some corn rakes and corn cakes and other food for the guests. Every family in the village offers their precious venison meat, rabbit meat, squirrel meat, etc. The host takes it home to barbecue, then chops meat and mixes it with steamed glutinous rice, and put it in a dustpan. This was the staple food of the banquet. The host takes out the self-brewed corn wine. At the beginning of the banquet, the host and guests slowly taste the wine and chatter. When the guests are stuffed, the host will hold the bamboo tube full of wine in both hands and walk to them and say: "You are like a star in the sky, coming from a distance, and will never disappear in our hearts. May our friendship be like torrents flowing continuously in the Nu River". Then holding a bamboo tube with the guests, the host and the guests hold each other's necks, and drink it together. This is called "bilateral pour", also known as "Tongxin wine", which is the peak of the banquet.

Section 2　Various Drinking customs

In human society, the way of eating any kind of food will be affected by the social structure and cultural characteristics, thus forming a rich and colorful food culture and customs. Among them, the drinking rules and customs are the most complicated and culturally characteristic.

Drink Fireplace wine

Fireplace wine, that is, drinking at the fireside and its related regulations. Fireplaces occupy an important part of ethnic minorities' life. In ethnic groups' areas, drinking at home is almost inseparable from fire pits.

In the traditional Yi society, the "top position" of the fireplace refers to the position of the door behind the wall. This position is closest to the altar and is reserved for male elders in the family. For Mosuo people of Naxi ethnic group, this position belongs to female elders, the head of the house. The male elders occupy the upper part of the fire pit in most ethnic groups, then followed by the female elders, the eldest son, the second son. In the past, the daughters and the daughters-in-law have almost no place at the side of the fire pit. Some just sit under it, specializing in adding firewood.

Sitting around the fireplace and drinking, the person who pours the wine for others is usually the eldest son of the family. The first cup is to be offered to the male elders, then the female elders, and the peers are filled in order of age. If a guest comes, the male elder will hold the pot and pour the wine for the first time. After it, the jug will be handed over to the eldest son, who will fill it up in turn. When drinking, one must first respect the guests or elders, clink glasses without toasting and drink more and less at will.

The fireplace wine of the Lisu, Nu, and Dulong peoples tends to create a relaxed, comfortable, warm and cheerful atmosphere. The wine song of the Dulong People goes like this: Our pigs are fattened, and our rice wine is cooked. Everyone is welcome to eat meat and drink around the fire pit. If you kill pigs at your house later, everyone will go to your house. This is our habit; this is our custom. Energize the fire in the fir epit, eat full, drink a lot, drink enough wine to sing "Pearl", and you will be energetic to work hard tomorrow.

Drink Za wine

Za wine is not a real wine, it is a special drinking custom. It uses bamboo pipes, rattan pipes, reed rods and other pipes to suck wine from a container into a cup, a bowl, or directly into the mouth. Because of the different straws, this way of drinking wine is also called bamboo tube wine, rattan tube wine, etc. It is popular in Yi, Bai, Miao, Lisu, Pumi, Wa, Hani, Naxi, Buyi, Zhuang, and Dong and other ethnic groups in Sichuan, Yunnan, Guizhou, Guangxi and other places. The wine drunk by this method are all water wines, including cold wine and hot wine. Cold one is to move out of the wine jar and insert a straw into the bottom of the jar to drink; hot one is to heat the water wine in a pot or directly put the wine jar on the fire, and drink while heating. The straws are inserted to the end, while drinking, add cold water to keep the wine in the jar or pot at the same level until the taste of the wine is lost.

This drinking method has long been popular among all ethnic groups in the southwest and is the highest etiquette for hospitality. XuXiake, a traveler in the Ming Dynasty, traveled in central Yunnan. This unique drinking style surprised him during dinner at the Tiejiachang villager's house on the Erhai Sea. The Dai society on both sides of the Yuanjiang River in southern Yunnan also has a similar drinking method. During the festive season, a barrel or a large tank filled with water wine is placed on a wide dam with several bamboo pipes inserting among them. People sing and dance around the wine tank, and when thirsty, they approach the wine tank and drink from the bamboo pipe, clear the throat and sing again; when tired, they get close to the barrel and take a breath, refreshing spirits and dancing again. Guests are welcome to join the singing and dancing on the spot. Everyone rushes to the wine barrel, the host invites the guests to insert the pipes into the jar, and then they have a drink together.

The Pumi, Wa and other people use bamboo pipes to suck out the wine and place it in gourds and bowls for drinking. SuliMa Tune is a traditional ballad that reflects the process of making SuliMa and the joy of enjoying wine. On the Dragon Boat Festival in May, I went up the mountain to cut rock-gold bamboo, and made it into a pipe to suck wine. The pipe is like a curved willow branch, taking out Sulima in the earthen jar. The mellow Sulima was poured into a silver bowl with gold rim, and the wine thickened with bubbles. Grandpa and Dad raised their bowls with smile and drank with a sip."

Drink Turning-wine

Drinking turning- wine is a form of drinking in which a group of people sit together, share a jug or a bowl, and pass the same drinking in a certain way. It is popular in Yunnan, Sichuan and other places. Roaming in the minority areas of southwestern China, you can often see in fields, roadsides, street markets, groups of three or five people sitting around, in the middle there is a jug and an earthen bowl for wine, everyone passes a jug or bowl in a certain direction, takes a sip, and tells a joke, laughs and breaks up until the jug is empty, especially on the day of going to the market.

It is commonplace for Yi, Lisu, Miao, Nu and others to drink turning-wine. The Yi folk song circulated in the Ailao mountainous area sings the essence of drinking turning-wine.

"The ancestors of the Yi family have been open-minded since ancient times. We like liquor, and we respect our distinguished guests. No matter where you are from, no matter how poor you are, as long as you walk into Yi mountain, we are a family; as long as you are sincere, we are friends! Even if we are too poor to wear a coat, even if there is only half a spoon of rice, even if there is only half a chicken, even if there is only half a piece of pie, even if there is only one sip of wine, we all have to eat half of it, and we all have to share one sip. Because you are a distinguished guest, because you are a friend of the Yi family, because we are the descendants of an ancestor, we are originally leaves on a tree."

According to the legend, there is a big mountain on which the three ethnic groups of Han, Tibetan and Yi live. They worship as brothers, Han is the eldest brother, Tibetan is the second brother, and the Yi is the third young brother. They are all reunited during the holidays. One year, the third Yi brother harvested much buckwheat, and cooked a lot of noodle. Two elder brothers are invited to come and share. They didn't finish the meal on the first day, and a strong aroma of wine came out on the second day. After scooping it into the bowl, the three brothers pushed one another to drink the wine, from morning to night, they didn't finish it. Later, the gods told that as long as they worked hard, there would be new ones, so the three people took turns to drink, and they all got drunk.

Drinking the turning-wine is also a way of hospitality for Miao people, it is often drunk during the holidays or in warm atmosphere. The host poured half a bowl of wine in front of each guest, and everyone passed their wine to the person on the left, asking the next one to pick up the wine with their right hand. People gathered in a circle, and to show respect, the eldest one in the circle drank first. Then everyone drank in order for a round; then the host poured the second round of wine, and after drinking up, another round is started. This goes back and forth until one is drunk.

Drink Hands-over wine and crossed-arms wine

There are different forms of toasting during the banquets of Zhuang people, such as drinking hands-over wine, crossed-arm wine and turning-wine. When the host and guest are toasting, the guest drinks the wine from the host's cup, and the host drinks the wine from the guest's cup. Both the host and the guest hold the wine with their right hands, their arms cross each other, and each drink the wine in their own cup, which is called crossed-arm wine. The host and the guests sit around the table and toast each other at the same time. Each person drinks the wine from the cups of relatives and friends next to him, which is called drinking turning-wine. When the Zhuang people drink hands-over wine, they have to fill the cup and drink it all at once.

Drinking hands-over wine is also popular in the Miao and Shui ethnic regions. There are three ways to drink, the first two are similar to the above-mentioned methods, and the third is to exchange wine cups and bowls for drinking. You drink my wine, and I drink your wine, to express the sincere and heart-for-heart friendship. The Shui People takes wine as a gift, and consider it a precious one. When drinking to the climax, the host suggests that everyone raise their glasses, which means to drink reunion, hands-over wine. So everyone takes the wine glass sent by others with their left hand, and passes the wine glass to others with their right hand. When everyone raises their cups hand in hand, the oldest person in the middle of the distinguished guest drinks first, and then everyone drinks in from left to right. When everyone toasts and drinks, the rest of them must cheer for him, shouting "Xiu, Xiu" in unison, which means a toast.

Drink Gateway wine

Gateway wine is a way to greet the guests by setting up a series of passes on the road of welcoming guests, which shows the sincerity and determination of the host's hospitality. The guests must drink before they can pass the gateways. It is popular in southern Miao, Yao, Dong and Buyi ethnic groups.

The Yao people are addicted to wine and famous for their hospitality. They have the custom of drinking "Three-Gateway Wine". Whenever there is a celebration, the guest arrives, the suona and crackers are sounded, and the host takes the wine to form three passes outside the house. For each pass, the guests are offered with two cups of wine, which is called "three passes to welcome guests with six cups of wine".

Road-blocking wine is a custom of welcoming guests of the Miao people. When guests enter the village, people will prepare for road-blocking wine on the road in front of the door, sing road-blocking songs to the guests, and let the guests drink the wine. The number of road-blocking wine varies, ranging from three to five to as many as 12, and the last one is located at the entrance of the village. Guests have to drink one by one before entering the gate, which not only expresses the hosts' enthusiasm for welcoming the guests, but also the hospitality of the host.

Drink potluck wine

Drinking potluck wine is a custom shared by many ethnic groups. It is a form of shared-drinking by many participants. The drinking and dancing mostly happen in the spring and flowers-blooming season, the locations are mostly in the woods and grasslands far away from the village, most participants are young boys and girls in the blooming season. Potluck wine is also called "eat mountain wine" and "drink flower wine". Festivals such as the Tanhua Mountain Flower Arrangement Festival of the Yi people in Dayao County of Yunnan Province, the February 8th Festival in Weibao Mountain of Weishan County, the Flower Mountain Festival of the Nu People in Gongshan County, and the Miao Mountain Flower Festival in Southeastern Yunnan, all have a direct relationship with the drinking customs of potluck wine.

"Go and climb the Western Hills on March 3rd". On the banks of the 500-mile Dianchi Lake, the "Flower Wine Party" of Biji Mountain and the "Drinking Flower-Wine " of YuAn Mountain have a long history and a wide range of influences, and they are still popular today. The Yuan Wine Party is said to be created by Duan Suxing, the king of Dali. He once built palaces in Kunming, he also built Chundeng Dike and Yunjin Dike, and at the same time he planted a wide range of exotic flowers and plants. In the warm spring season, young men and women were invited to travel around Yuan Mountain with wine and food, and he then built ditches to lead water to make nine-curve bridges, with small wooden boards floating on the water, and glasses put on the wooden boards. When wine boat coming and stopping by one person, they should drink wine and have fun with grass hairpins.

Drink Tongxin wine

Two people share the same drinking vessel, either standing side by side, sitting side by side, kneeling side by side, or squatting side by side, then two people fold their shoulders and neck, ear to ear and face to face, one with left hand and the other with right hand, hold the cup (tube or bowl) at the same time, put their mouths together, and drink at the same time, which is called drinking Tongxin wine. The wine can be drunk up, or take a sip, rap for a while, and then sip again, and repeating until the wine is exhausted. The wine utensils include wooden bowls, bamboo tubes, horn cups, croissant cups, trotters cups, etc. This drinking method is available in many ethnic groups, which is most common among the Yi, Miao, Lisu, Nu, Dulong and other people.

Lisu Tongxin wine embodies the enthusiasm and boldness of the Lisu nationality. The Lisu people live in scattered, high mountains and large rivers for a long time. Once they get together, they drink wine, sing and dance to express their affection, especially using drinking methods such as "three cups of wine" and "double-cup falling wine" to express their emotions. Lisu people call this colorful and unique way of drinking "Tongxin wine". Tongxin wine shows Lisu people's affection for relatives and friends, and it also shows that they are of one heart and one mind. There are six different ways to drink Tongxin wine with different meanings.

The first type is called "Yahabazhi" (Stone Moon Wine). In the depths of the Nu River Grand Canyon, there is a natural rock cave in Gaoligong Mountain in Lishadi Township, Fugong County, which is like a bright moon hanging over the sky. The Lisu language calls the stone moon "Yahaba", which is in the hearts of all Lisu people. It is also the birthplace of the Lisu people's ancestors and their roots. Ahabazhi embodies the Lisu people's pursuit of unity, respect for friends, purity and sincerity. When drinking, everyone stood around the table, holding a wine cup with their right hand, and at the same time holding friends or guests with their left hand, the whole scene was like a full moon. After singing the toast song, everyone said "Yilaxiu" (which means one toast).

The second type is called Sannizhi (Three-River Parallel-Flowing Wine). The Lisu people are the masters of the Jinsha River, Lancang River and Nu River. The main settlement area is in the core area of the "Three Parallel Rivers" scenic spot. The Lisu people regard the water of the three parallel rivers as good wine, and combine it with drinking, showing the beauty of Three Rivers' scenic spot, and the harmony between man and nature. When drinking "Sannizhi", three people put their left hands together and approach one other, and with the cup on their right hand twisted in a counterclockwise direction to form the Three Rivers, which symbolizes three people working together to create a better tomorrow.

The third type is called "Rankazhi" (Warrior liquor). Warrior liquor, also known as hero liquor, is the way Lisu warriors drink when they "go up to the sword mountain and go down to the sea of fire". Drinking this wine means having incomparable courage and determination to overcome all difficulties and obstacles. First, the elders or Nipa (priests) gave the warriors "farewell wine", and the toasters offered two cups of liquor to the warriors at the same time and said; "Nizizhiduo", the warriors gave thanks after drinking. The second is offered at the triumphant-return of the warrior. The elders or Nepalese serve two cups of liquor to the warriors. When drinking, the warrior first turns his head to the right in a half-squatting style. The warrior tilted his head to the left, and Nipa tilted his head to the right to signal affirmation, and toasted the wine in his right hand to the warrior, so that the warrior would become a hero who returned triumphantly.

The fourth type is called "Puhuazhi" (fortune wine). Puhuazhi is the long-cherished wish of the Lisu people in the struggle against nature for success and wealth. When drinking, the two of them crossed their hands on each other's wrists while holding the other's hands with theirs, and with the lower limbs crossing. They looked like an Arabic number "8" horizontally and vertically. The upper and lower two eights (8s) express the good wish of "making fortune". When drinking, both sides will speak "Puzhihuaduo".

The fifth type is called "Sijiazhi" (Missing Wine). Sijiazhi is the most common way of Lisu people drinking Tongxin wine, it also known as brothers' wine, and brother and sister wine. It's a way to express deep affection to friends, guests and relatives from afar. When drinking, two of them face each other with their right arms around each other's neck, and the left hand gently supports each other's back and say "Nichizhiduo" before drinking. The other is that two of them put their arms around their shoulders and face-to-face, with their mouths close together, and drink the wine up.

The sixth type is called Rishizhi (longevity wine). The Lisu people have a tradition of respecting the old and loving the young. There is a saying that respecting the elderly is more important than respecting the heaven and the earth. When paying respect to the elders, the juniors knelt holding the wine glasses in both hands, and bowed to the elders for three times.

Toast on horseback

In the long-term social and historical development, horses and mules are the most important means of transportation for mountain peoples. When the distinguished guests arriving, the host is waiting outside the door. Before the guests get off the horse, they will offer wine to toast immediately. Toast before the guest getting off the horse is the highest etiquette for the Yi, Bai, and Naxi people to receive distinguished guests, and it is mostly used in solemn and important occasions.

Nowadays, in cottages with roads, when important guests enter the village, people will gather on the sidewalks to welcome them. Before they getting off the car, the host will toast a bowl of clear and sweet wine. If you receive a toast in the car or on the horse, it shows that the cottage has regarded you as the most distinguished guest. You can't refuse the wine presented to your chest. You must pick up the wine with both hands, raise the cup to the eyebrows to show gratitude and reciprocity, and then drink it all. If you are really incapable of drinking a lot, you must bring the wine to your mouth and sip a little.

Sing and dance to persuade wine

Most ethnic groups are good at singing and dancing, and introducing ethnic singing and dancing into banquets to encourage drinking is another unique landscape of ethnic group wine culture.

The custom of singing and dancing to persuade one to drink is especially prevalent in Yi society. In the cheerful times, the banquet of the Yi village is composed of a small orchestra with a toast and persuasion team, one holding a pot, one holding a horn cup, one singing, one playing the gourd Sheng, and one playing the bamboo flute or xiao (a vertical bamboo flute), the mouth chords and harmonica. One person plays the three-string musical equipment, and dances to the distinguished guests. The pot holder retreats to the side after pouring wine, the cup holder raises the cups to his eyebrows. If the guests don't drink, singing and dancing continued. Guests should stand up, wait them to finish singing the song, then thank them with songs, take the wine with both hands and raise to their eyebrows, either drank them or sip lightly, then raise to their eyebrows with both hands again to give thanks, and also give a toast in return.

For the Yi people in Xiaojieshan District, Ershan Yi Autonomous County, Yunnan, during festive seasons or wedding celebrations, guests sit on the ground in a courtyard paved with pine leaves, drinking and chatting, and the host (usually male elders) leads a group of youthful young boys and girls to come and they gather the guests in the middle, and persuade them to drink one by one. The host holds wine by himself, a young man next to him holds a pot of wine to pour for others, a young girl takes the wine and raises a glass to wait, the host calmly sings: "Toast you a glass of wine, please have it, drink this wine, it is to express the heart of Yi people. Toast you two glasses of wine, please have it, drink these two glasses of wine, I wish you happiness. Toast you three glasses of wine, please have it, drink these three glasses of wine, I wish you good fortune."

Drink stalk wine

The traditional liquor of the Yi people is stalk wine. Filled with jars, and when drinking, people use thin bamboo stalks to insert into the jar to drink or use it to connect to the wine glass for drinking. Because the stalk wine is filled in jars, the "jar" is used as the counting unit. Nowadays, a bottle of wine is also called a "jar" of wine.

Another feature of drinking stalk wine is the use of the Sama (scale, mark) system. That is, drill a small eye on a bamboo slice and insert a small bamboo stick. When drinking, place the bamboo slice horizontally at the mouth of the wine jar with the small bamboo stick facing down to form a sama (equivalent to a glass of wine).

Stalk wine is a kind of water wine with a low alcohol content, generally between 20 and 30°. The wine is mellow and sweet, suitable for all ages to drink throughout the year. Every time you drink, pour the water into the jar and mix it with the wine until the water is level with the mouth of the jar. Drinkers must drink with stalks until Sama's small bamboo strips are completely exposed, as a toast to Sama. Fill up with water again, and the second drinker must drink the Sama bamboo sticks completely, and repeat this process until the wine tastes as light as water. A large jar of stalk wine can be drunk for several days, generally a jar of stalk wine is enough for the New Year.

Drink shout wine

The Miao nationality has a unique way of drinking, pulling ears to drink shout wine. The Miao family is not interested in playing finger-guessing game when drinking, but they drink shout wine to add to the fun. When drinking, they start to drink at will. After the three rounds of wine, the host stands up with the glass or bowl, and everyone stands up with the glasses. A passes his own wine to B, and B also passes his wine to C, in turn, and the last one exchanges his own wine to A. Everyone stands holding hands, bends over and shoulders by shoulders, forming a big circle. The host or one of them recites the wine theory, sings the wine song, and when it comes to the climax, everyone lets go of their voices and shout "yuoyuo!" After drinking a round of wine, they feed each other meat. When the wine is half full, they begin to pull their ears to drink. There are two ways: individual pulling, A and B exchange wine with each other, and both sides hold the wine glasses with their right hands and pass them to each other, and at the same time stretch out their left hands to pull each other's ears, and then drink up the wine that the other party has handed over. Collective pulling, everyone stands up and raises the wine glass. A passes the wine to B. At the same time, he stretches out his hand to grab B's ears. B passes the wine to C like this and pulls C's ears... After all the ears are pulled, everyone drinks the wine at the same time. After drinking, sometimes the meat is inserted into each other's mouth one by one, and then put down the hands.

Drinking Neck-Embracing Wine

Neck-embracing wine of Zhuang people can be carried out during a toast or when warmed with wine. When drinking, the two people stand side by side, you hold my neck, I hold your neck, and then each takes out a hand to hold the glass, you give me a drink, I give you a drink, and drink at the same time. Generally, only one-drink shows brotherhood.

Section 3　Drinking Taboo

Wine culture is an important part of ethnic group etiquette culture. All ethnic groups have some rules and taboos in pouring, toasting and persuading wine.

Pouring Taboo

It is a common custom of all ethnic groups to pour the full-cup wine for guests to show respect. The Yi nationality proverb says: "The full-cup wine is to show respect, and the full-cup tea is to bully one." The Hani ancient song goes like this: Take a big bowl when drinking, and pour it full. Take a big bowl when eating, and fetch the fattest piece of meat. When drinking, making the sound of drinking wine as good as a stream flowing; when taking dishes, and the chopsticks are to be as good-looking as butterflies feeding on flowers.

In ethnic group areas, whether in Yi mountains, Miao valleys or Zhuang and Dai villages, when relatives and friends gathering, the host prepares cup and bowl to treat guests with wine. When pouring, one must hold the pot with both hands and gently pour the wine into the cup until a full glass. Half pouring or uneven pouring will cause dissatisfaction among the guests. People who are good at pouring wine for others can use the surface tension of the liquid to slowly pour the wine into the cup and bowl, so that the wine slightly protrudes out of the cup and bowl without overflowing, which is to show respect to the guests. If there are no outside visitors, the environment for drinking at home should be relatively relaxed, and the person who pour the wine can moderately control the amount of drink by following the drinker's ideas.

The so-called concept ,"if the old don't go away, the new won't come", is very obvious in the wine-pouring custom. The Yi nationality proverb says that "if we don't drink up old wine, and we cannot make good new wine ". While Hani people say it more vividly, "The host's kind rice wine has whitened the face of the sky. The rice wine carried by the man is to be drunk a little from the bowl; the meal presented by the hostess is eaten a little from the bowl. The rice wine that the guests drank will return to the owner's wooden building with the smoke rising from the mushroom house; the meals that the guests ate will return to the Hani cottage with the white clouds rising in the mountains."

Toast Taboo

The Yi and Miao people often make a toast with horn cups, croissant cups or trotters cups when their guests arrive. You cannot bend the corner of the cup towards the guests when toasting, it will be considered unfriendly. When the guest responds, he should not bend the corner tip toward the host, but raise the corner cup with both hands, so that the corner tip bends to the right or toward him. When Dai people toast and drink, they have the custom of pouring a little wine on the ground. It is said that the devils smeared poison on the side of the wine bowl. After being discovered by a wise man, people set aside a little bit of wine before drinking to wash off the poison on the mouth of the bowl, so that they can drink the wine and avoid the harm. So when you visit Jingpo cottages, Dai and Wa mountains, if you see the host pouring a little on the ground when toasting or before drinking, don't be surprised. The behavior of the host is to show respect to the guests, and it is also a kind of self-protection in the long-term development of ethnic minorities.

Wa people also have many taboos to toast. In the Awa mountain area, when handing wine to the Awa people, stretch your hands forward with the palms facing upwards. Never keep your palms down and your thumbs separated from your four fingers. When drinking, the host puts a few drops of wine on the ground, takes a sip before handing it to the guests.

The guests take the bowl with both hands, and drink the wine without leaving the mouth from the bowl to show respect to the host; if they really can't drink, they should take a sip, apologize to the host repeatedly, and let others help to explain to the host. Another custom of drinking is that the host and the guest squat on the ground, the host passes the wine to the guest with his right hand, and the guest must also receive the wine with his right hand, and then he shows respect by pouring a little wine on the ground to the host's ancestors before drinking. For ethnic groups with the custom of drinking turning-wine, when sharing a bowl or tube for drinking, one should use the palm of to gently wipe the mouth of the bowl or tube after drinking, so the remaining wine on the side was wiped clean before handing it to the next person with both hands to show mutual respect.

Taboos of persuading wine

Jingpo people are warm and hospitable, and the host will warmly entertain all guests. Guests must use both hands to pick up the liquor and tobacco handed by the host. When acquaintances toast each other, they don't drink it after they receive it, but pour a little back into the other's wine barrel before drinking. The host thinks this is a sign of mutual respect. If several people go to Jingpo's house together, and the host usually doesn't toast one by one in person, but gives the wine barrel to elder ones, which means the host has given his heart to you, meaning that you can represent his heart and give everyone a toast. When drinking, people use the lid and don't drink directly from the canister. According to the custom of Jingpo people, visitors, whether male or female, the host will begin to toast them with cigarettes and wine as soon as the guests enter the door. Jingpo people's toasts have many characteristics. One is that when guests arrive, they use bamboo wine barrels to pour the wine to guests directly. One is to put the wine cup next to the fire pit and slowly offer it to the guests to enjoy. Another is to talk and toast to the guests at the same time. Whenever toasting the wine or tobacco, Jingpo people generally respect the elders first, then the peers, and then the younger generations.

The Buyi people also have the custom of singing to persuade the guests to drink, which includes ways like two-person duet, group duet, and folk song duet. At the banquet, the host will hold wine and sing the song of wine first: "The distinguished guests come to my house, like a phoenix falling on a deserted slope, like a dragon playing in shallow water. I'm afraid I have been a poor host. After listening to the toast song, the guests raise their glasses and sing a "praise song" in return: "Drink and sing a song of praise, and you sing and I answer, and wish the old man a longevity; blessing the boys to work hard; blessing the girls to weave smartly; blessing the host's family a good harvest every year." In this way, those who accompany guests have to drink it all. Every drink must match a song, and guests or those who accompany guests fail to respond will be punished to drink a "dumb cup wine". In some cases, a girl who can sing usually makes a toast to the guests. Every time the girl sings a song, the guest should also sing one back. Those who fail to sing back will be fined a glass of wine. Amid the cheers, the guests were drunk. When the guests said goodbye, and the host sings the farewell song, and sends them to the door. Everyone has developed deep friendship and they are reluctant to leave each other.

The Achang people pay much attention to alcohol ethics and think that being drunk is a shame. Whenever the hospitable host pours and persuades wine for many times, many wise guests express their gratitude to the host on the one hand, and on the other hand they use the folk saying of "wine in the pot, people talk about wine, wine in the belly, wine manipulates people" to state that he really can't drink anymore. When the host hears these words from the guests, he will usually stop persuading them to drink.

Chapter Six   Affectionate Toast Songs

The ethnic minorities in Yunnan are good at both making wine and singing. They use songs to narrate their ancient history, traditional lifestyle, customs, and their yearning for a better life. Wine is inseparable from their daily life, therefore, the custom of "good wine must be accompanied by songs" endows wine songs more cultural connotations which exhibit the temperaments, cultures and traditions of various nationalities.

Wine song is also called "wine etiquette" or "wine etiquette song". With a long history, wine songs have become the crystallization of the wisdom of all ethnic groups. The lyrics are improvised, rich in the rhetoric techniques such as pairing, simile, exaggeration, and parallelism. The creation and the singing of wine songs lead to the integration of wine and folk literature, which vividly reflects the unique lifestyles, customs, diligent and frugal virtues and beautiful hearts of ethnics. Currently, the popular ethnic wine etiquette songs are divided into brewing songs, toast songs, wedding songs, festival songs, sacrificial songs and so on. The duet between host and guest conveys not only emotions but also generosity, courtesy and hospitality.

Section 1   Wine Songs for Toast

All ethnic minorities in Yunnan are warm-hearted and hospitable. They entertain guests with mellow wine and sweet singing. The aroma of wine is intoxicating, and the sweetness of wine songs is nostalgic. All guests who come to ethnic minority villages will receive such a cordial invitation:

(Pumi people's wine song for welcoming guests)

Welcome friends,

My dear friends,

Come to take a seat on Pumi's flower bench.

Tea in wood bowls,

Milk in jade bowls,

Freshly brewed Sulima in bowls with golden rim,

Sweet honey in bowls with silver rim.

Pumi people are the most hospitable,

Come to take a seat on Pumi's flower bench.

(Lisu people's wine song for welcoming guests)

Please come to my village for a walk,

Please come to my house for a visit.

The sweetest Chu wine is here for you,

The most fragrant distilled wine is here for you.

Together we dance Qian Er, (Qian Er: Lisu Dance)

Together we sing You Ye. (You Ye: transliteration of Lisu language, a tune name of Lisu folk song)

Dance your happiness out,

Sing your happiness out.

Singing is indispensable when ethnic minorities drink. In addition to wine songs for welcoming guests, they also sing wine songs as they propose toasts, persuade guests to drink more and stay longer, or to express gratitude to host's hospitality. Buyi people love wine and are good at singing and dancing. Following the customs of "No etiquette is without wine" and "There must be songs if there is wine", holding the belief that wine and songs will show their sincere and enthusiastic hospitality, they would sing wine etiquette songs when entertaining guests. The host sings to welcome the guest and apologizes for the lack of hospitality, while the guest replies with songs, expressing heartfelt thanks for the host's hospitality, praising the wine, delicacies, and host's cooking skills.

(Buyi people's wine song for toast)

Host: Although this is a table for banquet,

There are only a bowl of taros

And a bowl of turnips on it.

Not a piece of meat adulterated,

No piece of fat mixed.

Guest: I eat at my home cabbage root,

While at your place, sea cucumber is the treat.

Wine is for intimate relations and meat for taste,

Thank you for your kindness.

I eat at my home chili,

While at your place, cubilose is the treat.

Wine is for intimate relations and meat for taste,

Thank you for your kindness.

Even if the table is full of delicacies, the host always sings like this to show modesty while the grateful guests will express their praise and appreciation to the host. The verse "Wine is for intimate relations and meat for taste" reflects the mediating role of wine in building and strengthening strong and affectionate friendships. Even though dishes are simple, the host's hospitality is better than the delicacies from mountains and seas.

At the Buyi banquet, every part of the agenda is a ceremony which is mainly performed through singing. When entering the house, the guests will sing Opening the Wealth Door to bless the host's wealth. Before the banquet starts, they will sing Setting the Table; having taken their seats, the song of Handing out Chopsticks; after chopsticks are divided, the song of Opening the Wine Pot, etc. .

Host: Pairs of bamboo chopsticks,

Let Mei[①] hand out to the female singer,

Let Mei hand out to the male singer,

Who joyfully sing on the stage.

Golden are the phyllotachy pubscens chopsticks,

Let Mei present to the singer,

Let Mei present to the singer,

---

① Mei, younger sister, a modest way for a woman to address herself.

Who merrily sing on the stage.

Guest: Pairs of bamboo chopsticks, neat and ordered.

Joyful am I

As sister handing out chopsticks to my hands.

We accompany each other at this banquet.

Golden are the phyllotachy pubscens chopsticks,

Handed out merrily

By smiling sister to my hands.

Let Mei sing to congratulate the host.

Don't Decline to Drink at the Banquet (Buyi people's wine song for toast)

Host: A table with four corners,

Seven dishes of leek and eight dishes of ginger,

All are plain and bland,

Without any broth.

It's difficult for guest soldiers to fight.

It is difficult to plant seedlings in the field without water.

Don't decline to drink.

I have no money for a sumptuous banquet,

But a sincere heart to receive my distinguished guest.

Guest: I'm visiting my sister, virtuous and bright,

Who offered several stewed and braised dishes,

And eight-treasure rice for every meal,

With abundant tea and fruits.

Being treated as a distinguished guest,

Remembering every bit of

My sister's cordiality and benevolence,

My heart would be satisfied even with a cup of water.

Zhuang people toast their guests with cross-cupped wine, using white porcelain spoons rather than wine glasses. Both host and guest will take a spoonful of wine from the wine bowl and drink in the cross-cupped way. In the mean time, the host will sing a wine song to propose a toast:

Filled with wine, the tin jug is white and shining.

Don't say No when the wine is presented,

With my sincere heart to distinguished guests,

Respecting you as a celestial.

Filled with wine, the tin jug and the white porcelain cup.

Don't say No when the wine is presented.

Not mellow, but is brewed with heart.

You become a celestial when drinking half a cup.

(Buyi people's wine song for toast)

Host: One cup of wine,

Pour will I.

Toast to our precious reunion,

However faint the wine will taste.

Guest: Two cups of wine, two full golden cups,

Which reflect the host.

Dear sister, you sincerely toast,

With effusion and passion.

Host: Three cups of wine, pour will I,

For guests to taste.

However plain is the wine, our friendship is permanent.

How rare is sister's visit to my place.

Guest: Four cups of wine, fragrant and sweet,

Full of your affection as deep as mountain valleys.

As the old saying goes, water is sweet

When there are benevolence and righteousness.

Host: Five cups of wine, pour will I,

To toast for dear kinsfolk.

Cups of bitter wine to express my goodwill.

How rare is sister's visit to my city.

Guest: Six cups of wine, fragrant and sweet,

Suffusing the singing stage and the city.

As the saying goes,

A courteous person is worth gold.

Host: Seven cups of wine, sour and tart,

Seven sisters gather for reunion.

Toast to my sisters,

How sweet are our hearts.

Guest: Eight cups of wine, fragrant and sweet,

Host pours and persuades guest to drink.

The Osmanthus wine brewed by our host,

Famous across the lakes and seas.

Host: Nine cups of wine, hot and spicy.

Pour the spicy wine for aunt,

And toast to sisters.

There is no good wine but good idea.

Guest: Ten cups of wine, ten full cups.

Ten sisters gather for reunion.

Your effusion and passion as deep as the sea,

Wealth and glory be yours for thousands of years.

(Miao people's wine song for toast)

Host: Knowledgeable in your advanced age,

A sincere toast to you, dear uncle.

Toast to youth for their love,

Toast to elders for their longevity.

Guest: Wasted time for decades,

Never been out of the village and have little knowledge.

Thank you girl for toasting me,

Wish you a happy marriage.

Yi people are not only adroit at brewing wine, but also hospitable. They always serve guests with mellow wine and beautiful wine songs, both when toasting or persuading guests to drink more, and when seeing guests off.

(Yi people's wine song for toast)

The first cup of wine, toast to the ancestors,

Who split the sky and the earth to stop the flood.

The second cup of wine, toast to the elders,

Who burned the weeds to grow buckwheat and raise their children.

The third cup of wine, toast to brothers,

The ancestor's motto must be kept in mind:

"Peacock looks better than golden pheasant,

Harmony is more precious than Phoenix. "

The first piece of meat, sacrificed to the victims,

Who passed away and left their families in tears.

The second lump of meat, dedicated to the poor,

Who worked hard day and night, sweating for years.

The third piece of meat, presented respectfully to my relatives,

"We two families share a wall,

Which is left by the ancestors. "

Do Drink the Wine, Like It or Not (Yi people's wine song for toast)

A Lao Biao[1], take up the bowl and drink.

A Biao Mei[2], take up the bowl and drink.

---

[1] Liao Biao, an intimate way to address male friends.

[2] Miao Mei, an intimate way to address female friends.

A Lao Biao, do drink whether you like it or not.

A Biao Mei, do drink whether you like it or not.

If you like, do drink.

If you don't like, do drink.

Whether you like it or not,

Do drink.

Whether you like it or not,

Do drink.

A Lao Biao, take up the bowl and drink.

A Biao Mei, take up the bowl and drink.

Brother has effusion, sister has passion.

Do drink until the moon sets.

Brother has effusion, sister has passion.

Do drink until the moon sets.

Drink enough rice wine (Yi people's wine song for persuading guests to stay longer)

Relatives, don't go, don't go,

There are so many pigs and sheep, there are so many.

Have killed nine small ones, nine small ones,

There are still ninety-nine, ninety-nine.

Friends, don't go, don't go,

There are so many crocks of rice wine, there are so many,

Have drunk nine small ones, nine small ones,

There are still ninety-nine, ninety-nine.

The food is delicious and tasty, delicious and tasty.

Drink enough rice wine, drink enough.

Friends and relatives sit together, sit together.

A toast to celebrate the harvest year, celebrate the harvest year.

Section 2   Wine Songs for Wedding

Many ethnic minorities in Yunnan have the custom of Yingqin[①] and singing wine songs at almost every part of the wedding celebrations, such as when proposing for marriage, marrying daughters off, weeping before marrying, sending the bride off, groom receiving his wife, and guest blessing the newlyweds. For example, at Buyi weddings, after enjoying wine, the elderly often sing "The Wine Song", "The Kaiqin Song" and "The Ancient Song" in the sitting room to congratulate the host who greets the guests with wine and tea .

When Yi girls get married, there is a custom of weeping before the wedding ceremony. Not only does the bride cry, but even the sisters who accompanied her would cry. In the meantime, people sing weeping songs, which sound passionate and touching. On the wedding day, the bride's families place a table in the sitting room. While the bride, her sisters and female cousins are sitting at the table, the family's daughters-in-law stand on both sides of the table and sing the wine song for persuading her to get married. In such sad singing and weeping, the bride is backed by her younger brother to circle around the table three times, then out of the sitting room. In the courtyard, beside the bonfire, all women continue to sing the song for marrying her off. At the same time, the bride's family will pour wine for guests so that all will drink while singing.

A procession of relatives and friends from the groom's side will come to escort the bride to the groom's house for the wedding ceremony. After the procession has arrived at the bride's house, the men in charge of sending the bride off (the bride sender from the bride's side) and the men in charge of receiving the bride (the bride receiver from the groom's side) will sing wine songs together right beside the gate of the house. The content of the wine songs ranges from the birth of mankind to the origin of the six ancestors, and so on. While singing, the bride and the woman in charge of sending the bride off (the female sender from the bride's side) are brought to the sitting room. After turning around three times from left to right, the bride starts to toast to distinguished guests. When toasting to the bride receiver, the bride kneels down in front of him and asks for ear bracelets. While drinking, the bride

① Yingqin: the groom and relatives go to the bride's home to escort her to the groom's home.

receiver will sing wine song to relate the origin and function of ear bracelets. After that, the procession escorts the bride to the groom's house.

As the procession arrives at the door of the groom's house, wine and delicious food have already been placed on a table in the courtyard. Both the bride receiver and sender will sing wine songs in an antiphonal style to indicate receiving the bride. The receiver sings first with invitation and questions, the sender follows with answers. After the antiphonal singing, the bride will acknowledge relatives from the groom's side by toasting to them. When toasting brothers-in-law, the bride must kneel down while holding the towel in her mouth and the wine in her hand. The bride sender would help her getting to her feet, then sing to entrust the bride to the groom's families, requesting brothers-in-law and their wives to help and teach the bride patiently. The bride receiver would respond with singing to accept the sender's request and express courtesy. Finally, the two gentlemen sing together to teach the bride about how to behave in the future on behalf of her parents.

After the bride's acknowledgement of the relatives and the toast, relatives set off firecrackers. Those who like to sing would sing together. They drink wine while singing to moisturizing their throats, while the joyous singing excites them to drink more. Singing, drinking, and singing, the jubilant and festive atmosphere reaches a new climax. That night, together with some singers, the host would sing and toast to the relatives who are temporarily staying with neighbors. They would sing the wine songs for toast first, then the wine songs for persuading guests to stay longer. Relatives will also sing to express their gratitude. On such a sweet sleepless night, while the bonfire is blazing merrily, the wedding is immersed in the sweet wine sings and the fragrance of wine.

Cu Mu Zuo Hou (excerpt from The Collection of Yi People's Wine Songs)

Female singer:　Dear two betrothal presents escorts,

　　　　　　　　Please don't be busy drinking!

　　　　　　　　When you came, there was not a small path in the mountains,

　　　　　　　　Did you grow wings and fly over the clouds?

　　　　　　　　The palace in the moon is shining,

　　　　　　　　Did the glowing beam pierce your eyes?

　　　　　　　　Did the thousands of silver needles pierce your tears?

The direction of the path is hard to discern as if it is painted with oil,

I wonder how many times did you fall?

The escorts: Dear singing Biao Mei[①],

When I came, it was a world of ice and snow,

Who dared to walk through the path,

Just followed the road and climbed the mountain ridges.

Ice and snow dressed up the mountains as a silver palace,

But no light reflected by the snow on a cloudy day.

Used to herding cattle and sheep on this road,

So I can recognize the direction with eyes closed.

Female singer: Dear two betrothal presents escorts,

Please don't be busy eating delicious dishes,

There are 297 sheep in the pen,

How many groups would you divide them into?

No group shall have more,

No group shall have less.

If you can't guess,

Three bowls of wine are the pass.

The escorts: Dear singing Biao Mei,

Open the shells we could see the nuts.

There are 297 sheep in the pen,

Divide them into three groups,

No group has more, no group has less,

Every flock of sheep is ninety-nine.

You lost, we won,

Will Biao Mei drink three bowls of wine!

"Cu Mu Zuo Hou" is an antiphonal song used by female singers from the bride's side to challenge the escorts during the wedding ceremony of Yi people. "Cu Mu" in Yi language

---

① Miao Mei, an intimate way to address female friends.

is the betrothal presents escort. "Zuo Hou" in Yi language is dining courtesy. The betrothal presents escorts are usually familiar with Yi's customs and are good at singing and dancing. As the representatives of the groom's family, the escorts must overcome all the challenges while escorting presents to the bride's house so as to glorify the groom's family. If defeated by the female singers while sing "Cu Mu Zuo Hou", as a punishment, the escorts would not be served with food, but go hungry.

At Miao weddings, the host and guests often invite singers to sing wine songs. The guests' singer sings in the sitting room, the host's singer sings in the kitchen, and others will take turns to drink while listening to the songs. Drinking and singing, singing and drinking, the host and guests sing in antiphonal style all night long.

(Miao people's wine song for wedding)

Everyone wants to find a girl who can do needlework,

Everyone wants to find a girl who can make mellow wine.

The girl who can do needlework comes from the place of sheep,

The girl who is good at brewing wine comes from the clouds.

She brings distiller's yeast from where the sun shines,

She brings distiller's yeast from where the moon is bright.

......

Stretch out your left hand for the wine bowl,

With your right hand, bring it to your mouth to drink.

Adults say the wine is fragrant and delicious,

Children say the wine is so fragrant and sweet.

Its reputation has spread far and wide,

Its good name is well known in the big cities!

The process of dating, proposal, and wedding of Pumi people are accompanied by songs. "Love Song" expresses the mutual affection of young people, while "The Tune of Renqin[①]" is sung on their blind date. On the wedding day, "The Tune of Yingqin" and "The Tune of Leaving" are sung when the groom's relatives and friends come to bride's home to escort

---

① Renqin: to acknowledge dating relationship.

her to the groom's home. When picking up the bride, people from the two sides will sing "The Tune of Marriage". "The Tune of Getting on Horse" is sung when the bride leaves home on horse; "The Tune of Getting off the Horse" when the bride meets with the groom halfway; "The Tune of Closing the Door" when the bride arrives at the groom's home; "The Tune of Welcoming Guests" and "The Tune of Being Guests" when relatives and guests congratulate the newlywed couple. All these songs are passionate and simple, as "The Tune of Jieqin[①]" sings:

You have come to the place where coral and agates pile up.

What did the elders say when you left home?

Corals and agates are the most beautiful decorations,

which I will bring home to dress up my bride.

You have come to the place where pearls are clustered.

What did neighbors say when you left?

Strings of pearls shine with dazzling light,

allow me to choose a string to hang on my bride's chest.

You have come to the place where silver is shining,

What did relatives say when you left?

Shiny silver is a valuable thing,

Allow me to carry a bag back to buy farmland.

You have come to the place where gold shines,

What did clans say when you left?

The shining gold is extremely precious,

Allow me to carry a bag back to build a house.

(Wa people's wine song for wedding)

Ah——

A girl of marriage age goes to her husband,

A boy of marriage age takes his wife.

---

① Jieqin: to receive the bride.

You two sharpened your long swords,

You two talked about love.

You two dated under the cotton tree,

You two courted under the eaves.

You two hunted and picked fruits together in the mountains,

You two fished and swam together in the rivers.

Having revealed your hearts to each other,

Having proved your honesty and loyalty,

A solemn pledge of love brought you to

A marriage of blooming flowers and full moon.

We have prepared Mang drum stick,

We have prepared the betrothal presents.

Now it is necessary to celebrate you marriage

And hold your wedding formally and orderly.

You will have daughters,

You will have sons.

Teach your daughters about business,

Teach your sons about reclamation.

From now on, you will live healthy and long.

From now on, you will live safe and sound.

A long life until your eyebrows bloom,

A long life until your beards turn gray.

Ah——Sai Xin[①]

---

① Ah —— Sai Xin: Wa language, meaning "Here is the wine. Here you go."

Section 3　Wine Songs for Festivals

There must be a banquet in the festival celebrations of ethnic minorities, and if there is wine, they would definitely sing wine songs. When festivals arrive, the villages of Pumi people are filled with excitement and jubilation. Pumi people sit around the fire pit, singing and drinking. You could hear pleasant singing everywhere. Their typical wine songs for the New Year festival include "The New Year Song", "New Year Greetings", "Worshiping the Gods" and so on. The tunes of the New Year wine songs are generally gentle, affectionate, and rhythmic, such as "The Spring Festival Song":

A li ya li ——[①]

The spring festival comes.

The green pine branches are sprouting,

The bright mountain flowers are blooming,

The spring has dyed the villages green.

Smoke from the kitchens are drifting above the wooden cottages,

Golden light dancing in the fire pit,

Around which happy people gather.

In this beautiful holiday,

We miss our distant ancestors.

Without past,

No present,

Without beginning,

No ending,

Without truth and faith,

Couldn't we tramp over the mountains.

Without industrious labor and bravery,

Couldn't we cross the abyss of poverty.

By relying on truth and faith,

---

① Ah Li Ya Li: a Pumi folk song supplementary word, which means "you sing and I sing".)

Could we reach the golden shore on the other side.

With hard work and courage,

Could we equip our Xining horses with golden saddles.

"Sulima" wine is fragrant and mellow,

You drink one bowl, I drink one bowl,

Its fragrance flows over mountains far away,

Drink to our heart's content and rejoice.

Pork belly is delicious,

One lump for you, one lump for me.

Sweat of hard work brings about delicacies,

Pumi people have abundant food.

Take one sip and warm the wine,

Take one lump and savour the meat,

The last words of ancestors are like strings of pearls,

Which melted in wine bowls,

And illuminated our future,

As bright and beautiful as silk and jade.

Day is past, twilight arrives,

Hard work brings about joy.

How jubilant the holiday night is,

With a joyous dance circle around the bonfire.

Hearts are united under the full moon,

While chanting the words of our ancestors.

To the flute sound, sweet and melodious,

Dance joyfully the sweethearts and friends.

Singing and dancing,

Pumi celebrates the New Year happily.

With blessings and song of praise,

Truth and faith are flying in sparks.

Fill the bowl with wine, with the jar with tea,

Dance jubilantly around the bonfire to welcome spring.....

Lisu's "Chuanqin1 Tune" takes the form of antiphonal chanting between "host" and "visitor", which vividly depicts the difficulties and obstacles encountered and overcome by the visitor, the yearning for meeting with relatives and friends, as well as the touching and beautiful reunion during the festival. At the end of the wine song, the visitor and the host sing respectively like this:

Visitor: Today is a beautiful day,

Tonight is an auspicious night,

We will meet,

We will definitely meet,

Meet with people who are expecting me eagerly,

Meet with people who treat friends sincerely.

You have spread the wool carpet,

You have arranged the bamboo stools,

Your wine is sweet and mellow,

Your dishes are fragrant and delicious,

You have served me with food and drink

That you have cherished and reserved.

I have eaten too much,

I have drunk too much.

Thank you, my dear cook.

Thank you, my dear wine brewer.

How much I have been missing you,

Too much to depict and account.

How many songs I have been singing while missing you,

---

① Chuanqin: visiting relatives and friends.

Too many to count.

It's time to turn around slowly,

It's time to go home leisurely.

Parting with you in tears,

Biding farewell in tears,

Don't like it, but have to part.

Don't want it, but have to separate.

I will visit my friends again if I am not too old to walk next year,

I will come back if I do not die next year.

Host: Our meeting is rare,

Our reunion is precious,

Why depart in such a hurry?

Why leave in such haste?

We live only one life in the world,

We have only one lifetime in the world,

Youth will not return,

Cannot be reborn after death.

How can you be willing to part so soon,

How can I be willing to separate so soon?

My heart is so sad,

My liver is so misery.

Can't hold back tears,

Can't hold back my sad tears.

I have been waiting for you year after year,

Like a lost bird.

I have been expecting you month after month,

Like a fallen leaf.

However sad you are, you have to part,

However misery we are, we have to separate.

Cuckoo sings by your side,

About how much I miss you.

Yang bird[1] sings near your ear,

About how much I yearn for you.

Hope you will visit next year,

Hope you will come next year.

New Year's greetings (festive wine song of Yi people)

Tonight,

We are going to celebrate the new year.

Where do months come from?

To count the years,

To count the months,

How to calculate the years and the months?

There are trees on the top of the cliff,

There are vines at the foot of the cliff,

The beginning and the end of a year have their dates,

As the trees have their numbers,

And the vines have their numbers.

The annual flower[2] is the flower of prosperous harvest,

The monthly flower[3] is the flower of flourishing livestock.

Every year when the grains are ripe in the autumn, many ethnic minorities have the custom of tasting new grains and drinking new rice wine which is commonly known as "Xingu liquor". After the harvest, Buyi people start brewing new rice wine and curing bacon. In the following year, when they are free from farm work during the first lunar month, they would invite relatives and friends to celebrate the harvest, drinking and singing.

---

① Yang bird: cuckoo.

② Annual flower: year.

③ Monthly flower: month.

On this day, the host family would get up early, prepare all kinds of delicacies, and wait for the arrival of guests. After the guests have arrived one after another, they are invited to sit around the fire pit which illuminates the room warmly. The host would propose a toast and sing "The Song of Modesty" with joy:

Host: The fire crackled last night,

The magpies twittered this morning,

Betokening the arrival of distinguished guests,

Who do come.

The host would open the wine jar. As the aroma of the wine overflows, the guests would sing.

Guests: In the middle is the rice wine newly brewed,

Whose aroma flows ten miles away.

If you wash the wine jars in the river,

The old dragon king would be drunk.

Host (humbly): Beautiful peacocks,

Come to rest in the thorny forest.

No food to eat when hungry,

No water to drink when thirsty.

I am neglecting the distinguished guests,

Who could enjoy nothing but sit.

Guests praise: Plenty of wine and meat on the table,

Eight bowls and nine jars,

Bending the table top,

Squashing the table feet.

In this way, the host and the guests congratulate each other and sing to the end of the banquet. When seeing the guests off, they would sing together at the edge of the village:

The winter plum blossoms are gone,

The peach blossoms are in bud.

Having enjoyed harvest wine,

Now break up and return home.

The spring breeze blows peacefully,

warming the land gently.

Every family chooses good seeds,

Every household fertilizes the field.

Cuckoo's singing is urging,

Waste not the planting season.

Plant early in the spring,

Grains pile up in the golden autumn.

Fresh rice wine every year,

Stout pigs and sheep every year.

Peace and health throughout lifetime,

Wealth and honor for generations to come.

After singing, the host and the guests bid farewell to each other, and agree to meet again in the coming year to drink "harvest wine".

Section 4   Wine Songs for Sacrifice

The ethnic minorities in Yunnan not only use wine for various sacrificial activities, but also combine wine songs with sacrificial rituals and sing sacrificial wine songs. According to the sacrificial rituals, the wine songs for sacrifice are divided into ancestor worship songs, exorcism songs, hunting sacrificial songs, farming sacrificial songs, festival sacrificial songs, and daily life sacrificial songs.

The Kucong people living in the Ailao Mountains have been hunting for generations. They call the hunting god "Shani". Every time they go out hunting, the hunters must kill chickens to sacrifice to the hunting god and sing "The Hunting God Tune":

Today, I sacrifice chicken and burn incense to worship you,

Cook rice for you and kowtow in front of you.

From now on, help me climb the mountains and go down the valleys,

Help me cross the rivers and go through the hurdles,

Allow me to rest at night and hunt beasts in the day.

Bestow me abundant food to eat all year round.

Help me detect beasts so that I don't shoot in vain, or waste arrows.

Protect me from beasts.

The wild boar is dedicated to you, the mantiacus is dedicated to you,

And the pheasants are sacrificed for you.

Whatever games I obtain,

All dedicated to you.

Sacrifice to the Hunting God (Yi)

Father the heaven, please drink!

Mother the earth, please drink!

Pi Wutuo, please drink! (The god who opened up the sky in the legend of Yi people)

Lie Zheshe, please drink! (The god who reclaimed the land in the legend of Yi people)

The three officials of the waters, please drink! (General term for various gods in the waters)

Prophetic gods, please drink!

Knowledgeable gods, please drink!

The mountain god in the mountain, the rock god on the rock, please drink!

The white rock god on the white rock, the red rock god on the red rock, please drink!

The tree gods at the top, side and foot of the mountain, please drink!

The cave gods at the top, side and foot of the mountain, please drink!

The rhino god at the rock peak,

The green snake god on the top of the mountain,

The gray snake god on the mountainside ,

The patterned snake god at the foot of the mountain,

The fish god in water,

The dragon god under water,

The four-legged lizard elf on the roadside,

The patterned fish elf under the bridge,

The river deer elf in the valley,

The mantiacus elf in the forest,

All gods, please drink together!

Will please the god of the heaven take charge of the sun so that it rises to dispel the clouds and casts bright light on the earth;

Will please the god of the heaven take charge of the moon so that it will be bright and the stars twinkle.

Will pleas the god of the earth take charge of the plants so that neither trees will shelter birds of prey, nor will bushes harbour beasts.

Sacrifice to all gods who bless hunters safe steps and auspicious life!

Sacrifice to all gods!

Please search the beasts and lead them out of their caves.

Please collect the birds and lead them out of the forest.

Bless me that I would be a great archer and never miss my target!

Bless me that all my prayers come true!

Wa people offer sacrifices to "Muyiji" before spring sowing. Muyiji, supreme god in the primitive religion of the Wa nationality, is in charge of wind and rain, life and death, and harvest. Wa people sacrifice pigs and oxen while singing sacrificial songs to pray for the blessing from the supreme god:

The god of the village,

The river spirit of the village.

Bailu flowers are blooming on the hillside,

Erythrina blossoms are blooming around the village.

The wind is whistling in the mountain,

The sun is blazing like a ball of fire.

It's time for us to plant seeds,

It's our planting season.

We are going to sow millet seeds,

We are going to plant seedlings.

It's the seedlings brought by the white pheasant,

It's the millet seeds brought by the titmouse.

Let them fall to the ground,

Let them go into the soil.

Let them take root even on steep slopes,

Let them sprout even on the rocks.

Please cover the beaks of birds,

Please cover the mouths of squirrels,

So that the seeds grow tall and strong,

So that the plants bend with plump ears.

We have prepared the warehouse,

May we have a bumper harvest.

Farming Sacrifice Song (Lahu)

The goose-tail flowers bloomed by the river,

The sparrow-tail flower bloomed in the mountain.

Brother picked a bunch for the mountain god,

Sister picked a bunch for the land god,

Bless us with exuberant rice flower,

Bless us with plump rice ears.

Bless the reclaimed wasteland with bumper harvest,

Bless the land with prosperous gains.

Not eaten up by birds or insects, not washed away by flood,

But abundant for people and livestock,

And overflowing the barns.

Sacrifice to Ancestors (Yi)

The deceased has three souls.

One goes with the ancestor,

So enshrine it on the incense table.

Make ancestor's body with grass,

Hands and feet with azaleas,

And bones with bamboo.

Paint it yellow,

Consecrate it in the shrine.

The sons come to offer wine,

The daughters come to offer meals.

.....

Your children and grandchildren,

Burn and offer sacrifices.

Ancestral spirits, come to eat!

Ancestral spirits, come to drink!

Worship ancestral spirits every year,

For blessings of peace and auspiciousness to your descendants,

For blessings of prosperity to livestock,

And abundance to grains.

"The Lusheng Lyrics" of the Miao nationality is sung to clear a path to heaven for the deceased. But almost every sentence contains the word of wine, which indicates that wine is essential for seeing the dead off. Miao people believe that, if the deceased have no wine to drink, their souls will not be comforted and they will not leave safely. Therefore, seeing off the dead is done by toasting.

Ah... the deceased, whom are you waiting for now?

Waiting for your children to come to toast?

They have toasted you seven times.

The deceased! Leave with the jar of wine if you can't finish it now,

You will honor your ancestors with the wine.

The deceased! Carry with you the wine toasted by families and friends,

And leave without hesitation!

You are waiting for toasts from families.

The deceased! Get to your feet and leave!

Do you have to wait until all relatives come to see you off by holding your hand?

Now, your family members have toasted nine times to you.

Take with you whatever you can't finish,

And offer it to our ancestors.

Summon the spirit of the dead (Kucong people of the Lahu ethnic group)

Hey, Zhale!

Hey, Zhale!

Grind rice, grind enough rice, grind enough for cooking.

Cook vegetables, cook enough.

Whatever you make, let it be more than enough;

Whatever you do, let it go as you please.

In the past days,

Wine made by me was not mellow;

Rice cooked by me was not enough;

Sticky-rice cake made by me was not delicious.

Today, the spirit of the rice and the spirit of water come home.

Make wine, make mellow wine.

Cook rice, cook abundant rice.

Pound sticky-rice cake, pound delicious cake.

Cook vegetables, cook enough vegetables.

Sift flour, sift enough.

Sift rice, sift enough.

Be a spirit and go home.

The water spirit and the rice spirit come home together.